The Seiberg-Witten Equations and Applications to the Topology of Smooth Four-Manifolds

The Seiberg-Witten Equations and Applications to the Topology of Smooth Four-Manifolds

by

John W. Morgan

Mathematical Notes 44

PRINCETON UNIVERSITY PRESS

PRINCETON, NEW JERSEY
1996

Princeton University Press books are printed on acid-free paper
and meet the guidelines for permanence and durability of the
Committee on Production Guidelines for Book Longevity
of the Council on Library Resources

The Princeton Mathematical Notes are edited by
Luis A. Caffarelli, John N. Mather, and Elias M. Stein

Library of Congress Cataloging-in-Publication Data
Morgan, John W., 1946–
The Seiberg-Witten equations and applications to the topology of smooth
four-manifolds / by John W. Morgan.
p. cm. — (Mathematical notes ; 44)
Includes bibliographical references.
ISBN 0-691-02597-5 (pb : alk. paper)
1. Four-manifolds (Topology) 2. Seiberg-Witten invariants.
3. Mathematical physics. I. Title.
II. Series : Mathematical notes (Princeton University Press) ; 44
QA613.2.M67 1995
514'.2—dc20   95-43748

The publisher would like to acknowledge the author of this volume
for providing the camera-ready copy from
which this book was printed

http://pup.princeton.edu

Printed in the United States of America

3 5 7 9 10 8 6 4

Contents

The Seiberg-Witten Equations and Applications to the Topology of Smooth Four-Manifolds

Chapter 1

Introduction

Beginning with the groundbreaking work of Donaldson in about 1980 it became clear that gauge-theoretic invariants of principal bundles and connections were an important tool in the study of smooth four-dimensional manifolds. Donaldson showed the importance of the moduli space of anti-self-dual connections. The next fifteen years saw an explosion of work in this area leading to computations of Donaldson polynomial invariants for a wide class of four-dimensional manifolds, especially algebraic surfaces. These computations yielded many powerful topological consequences including, for example, the diffeomorphism classification of elliptic surfaces. For some of the results obtained by using these techniques see [1], [6], and [2].

Last fall, motivated by new work in quantum field theory, Seiberg and Witten [9], introduced a different gauge-theoretic invariant which they claimed should be closely related to Donaldson's invariants. Indeed they gave an explicit formula for the relationship of their invariant to Donaldson's. While the link claimed by Seiberg-Witten between their invariants and Donaldson's has not yet been established mathematically, it is clearly true and can be shown to hold in all computed examples. Nevertheless, one can forget this supposed link and work directly with the new invariants as a substitute for the anti-self-dual invariants. This has been the approach during the last year or so.

It was clear from the beginning that the new invariants would be easier to work with since they involved principal bundles with structure group the circle instead of the non-abelian groups such as $SU(2)$ which arise in Donaldson theory. The surprise was that such simple invariants could capture the subtlety that Donaldson's invariants revealed. But in short

1

order, Witten [17] and then Taubes, Kronheimer, and Mrowka showed
that indeed these new invariants did capture these subtleties and that they
were easier to compute, at least in many cases. They did this by explicitly
solving the Seiberg-Witten equations over Kähler surfaces. (See [3] for
one account of the results for Kähler surfaces.) There followed in quick
succession a series of remarkable theorems, each extending in a different way
partial results from Donaldson's anti-self-dual theory. In fact, conjectures
which seemed reasonable from the perspective of Donaldson theory but
technically difficult, if not unreachable within that theory, suddenly became
the standard test cases for the power of the new invariants. One by one
these conjectures were established – leaving only one classical conjecture
outstanding. The remaining one, called the 11/8ths-Conjecture, deals with
the quadratic forms that arise as intersection forms of simply connected
spin four-manifolds.

It is the purpose of these notes to lay the groundwork for the Seiberg-
Witten theory and then to show how one computes these invariants for
most Kähler surfaces. We begin with the basics of Clifford algebras, spin
structures and their cousins $Spin^c$-structures, spin representations, and
spinor bundles. We then consider the Dirac operator on the spin bundles
over a riemannian four-manifold. The connection with Kähler geometry is
facilitated because of the close connection on a Kähler manifold between
the Dirac operator and $\bar{\partial}$. All of this is examined in detail. The book of
Lawson-Michelson [7] gives a complete introduction to this material as well
as an elaborate treatment of many more topics in the spinor geometry.

We then exhibit the Seiberg-Witten equations and show how to use
these equations to produce a finite dimensional manifold, the moduli space
of solutions to these equations modulo changes of gauge, inside an infinite
dimensional configuration space. The homology class of this moduli space
is then the value of the Seiberg-Witten invariant. We establish analogues
of familiar theorems from anti-self-dual theory. The equations defining the
moduli space are elliptic modulo the group of changes of gauge. A generic
perturbation of the equations leads to a smooth orientable moduli space
whose dimension can be computed by the Atiyah-Singer index theorem.
To orient this moduli space it suffices to choose an orientation of H_+^2 and
H^1 of the underlying four-manifold. As long as $b_2^+ > 1$ the moduli space
varies smoothly as we vary the metric and perturbation. When $b_2^+ = 1$
the generic perturbation yields a smooth moduli space, but there will be
singularities when we vary the metric and perturbation. These singularities
are reducible solutions to the equations. This leads to a chamber structure
for the Seiberg-Witten invariants in case $b_2^+ = 1$, similar to what happens

in the anti-self-dual theory.

We go on to establish one special property of this theory, namely the compactness of the moduli space of solutions. This is a consequence of *a priori* bounds for the pointwise norms of solutions to the equations. This result has no analogue for the anti-self-dual equations. Much of the geometric richness and much of the technical complexity of the anti-self-dual theory is directly related to the non-compactness of the moduli space. For both better and worse, that is all missing here. The Seiberg-Witten moduli spaces are compact and vary by a compact bordism as we vary the metric.

Having presented all these technical results, it is now clear that the homology class of the moduli space in the ambient space of configurations is an invariant of the underlying smooth four-manifold and the isomorphism class of the $Spin^c$ structure on that manifold. By definition, the Seiberg-Witten invariant of the $Spin^c$ structure is this homology class. We finish by explicitly computing the moduli spaces of solutions on 'most' algebraic surfaces, leading immediately to a computation of the Seiberg-Witten invariant for any $Spin^c$ structure over the surface. There are some special cases that we do not treat. These can, however, be treated by an elaboration of the techniques that we introduce. A complete discussion is contained in [3].

There is much more to be said about the Seiberg-Witten equations and the invariants that are defined from these equations. There are analogues of the theorems for Kähler manifolds which hold for symplectic manifolds, see [12, 13, 14]. There are also gluing theorems which lead to Meyer-Vietoris principals for solutions to the equations, see [8]. These then lead to product formulas for the invariants, which is one approach to gaining a more topological understanding of the invariants. In spite of the striking progress over the last year, much remains to be done. It is my hope that these notes will serve as an introduction making it possible for more mathematicians to contribute to this progress.

These notes are a written version of lecture series given at Columbia University and Princeton University during the past year. I wish to thank the graduate students at both Universities who attended these lectures and who, by their comments and questions, helped shape these notes.

Chapter 2

Clifford Algebras and Spin Groups

For any $n > 2$ the orthogonal group $SO(n)$ has fundamental group $\mathbf{Z}/2\mathbf{Z}$ and hence has a universal covering group called $Spin(n)$ which is a non-trivial double covering. There is a beautiful algebraic construction which yields the Spin groups (and much more as well). This is the subject of Clifford algebras.

2.1 The Clifford Algebras

An example. Before delving into the complexities of Clifford algebras in general, let us consider a simple example. Consider the unit sphere S^3 inside the quaternion algebra $\mathbf{H}$. Multiplication of quaternions induces a group structure on S^3. Let us consider the action of this group on $\mathbf{H}$ by conjugation

$$S^3 \times \mathbf{H} \to \mathbf{H}$$

$$(\alpha, \lambda) \mapsto \alpha\lambda\alpha^{-1}.$$

This action preserves the norm, i.e., is an orthogonal action. It also leaves invariant the center of $\mathbf{H}$ which is $\mathbf{R} \subset \mathbf{H}$, and hence it leaves invariant the perpendicular complement to $\mathbf{R}$ which is the three-dimenisonal space $\mathbf{Im}\,\mathbf{H}$ of purely imaginary quaternions. Of course, $\mathbf{Im}\,\mathbf{H}$ is naturally identified with the Lie algebra of S^3 and the action we are considering is the adjoint action of S^3 on its Lie algebra.

It is an easy geometric exercise to see that a unit quaternion α, different from ± 1, acts by conjugation on $\mathbf{H}$ preserving the complex plane $\mathbf{C}\alpha$ spanned by α and its perpendicular $\mathbf{C}\alpha j$. On the first complex plane conjugation by α acts trivially and on the second it acts by rotation through twice the angle θ between α and 1. It follows that the conjugation action of α on $\operatorname{Im} \mathbf{H}$ leaves invariant the line tangent to the circle generated by α and acts by rotation through 2θ on the perpendicular complement and thus that every rotation of $\mathbf{Im \, H}$ is in the image of the representation

$$S^3 \to SO(\mathbf{Im \, H}) = SO(3).$$

It is also clear that the kernel of this representation is the intersection of S^3 with the center of $\mathbf{H}$, which is the center of S^3 and is $\{\pm 1\}$.

In this way we construct the double covering group of $SO(3)$ and identify it with the group of quaternions of length one under multiplication.

The definition of the Clifford algebra associated to an positive-definite inner product space. Now let us turn to the general case. Let V be a finite dimensional vector space over $\mathbf{R}$ with a positive definite inner product $\langle \cdot, \cdot \rangle$ leading to a norm denoted $\| \cdot \|$. We consider the tensor algebra

$$T(V) = \oplus_{n \geq 0} \underbrace{V \otimes \cdots \otimes V}_{n \text{ times}}$$

generated by V. It is an associative algebra with unit 1. The Clifford algebra $Cl(V)$ generated by V is the quotient of $T(V)$ by the two-sided ideal generated by all elements of the form

$$v \otimes v + \|v\|^2 1$$

for $v \in V$. Notice that the grading on $T(V)$ descends to a $\mathbf{Z}/2\mathbf{Z}$ grading on $Cl(V)$, giving a decomposition

$$Cl(V) = Cl_0(V) \oplus Cl_1(V)$$

where $Cl_0(V)$ is a subalgebra and $Cl_1(V)$ is a module over this subalgebra. Corresponding to this splitting, we can write any $v \in Cl(V)$ as $v_0 + v_1$. We denote by

$$\epsilon \colon Cl(V) \to Cl(V)$$

the algebra homomorphism that is multiplication by $+1$ on $Cl_0(V)$ and is multiplication by -1 on $Cl_1(V)$; i.e., $\epsilon(v_0 + v_1) = v_0 - v_1$.

Choosing an orthonormal basis $\{e_1, \ldots, e_n\}$ for V we can write $Cl(V)$ in terms of generators and relations. Namely, $Cl(V)$ is the algebra over $\mathbf{R}$ generated by $\{e_1, \ldots, e_n\}$ subject to the relations $e_i^2 = -1$ for all $i \leq n$ and $e_i \cdot e_j = -e_j \cdot e_i$ for all $i \neq j$. In particular, it follows that every element of $Cl(V)$ can be written uniquely as a sum of products of the form

$$e_{i_1} \cdots e_{i_t}$$

where $i_1 < \cdots < i_t$. Thus, the dimension of $Cl(V)$ as a real vector space is 2^d where d is the dimension of V over $\mathbf{R}$.

Examples. (i) Let $\mathbf{R}^n$ denote the usual Euclidean space of dimension n. Then $Cl(\mathbf{R}^1) = \mathbf{R}[x]/(x^2 + 1) \cong \mathbf{C}$. The subalgebra $Cl_0(\mathbf{R}^1)$ is identified with the reals and $Cl_1(\mathbf{R}^1)$ with the purely imaginary complex numbers.

(ii) Similarly, $Cl(\mathbf{R}^2)$ is the algebra generated by x, y subject to the relations

$$x^2 = -1; y^2 = -1; xy = -yx.$$

Hence, $Cl(\mathbf{R}^2)$ is isomorphic to the quaternion algebra $\mathbf{H}$. The subalgebra $Cl_0(\mathbf{R}^2)$ is generated by xy and can be identified with $\mathbf{C} \subset \mathbf{H}$.

(iii) $Cl(\mathbf{R}^3)$ is of dimension 8 over $\mathbf{R}$. It is generated by x, y, z with $x^2 = y^2 = z^2 = -1$ and $xy = -yx, xz = -zx, yz = -zy$. This algebra is isomorphic to $\mathbf{H} \oplus \mathbf{H}$. The isomorphism between $\mathbf{H}$ and the first factor, resp., the second factor, is given by sending $1, i, j, k$ to

$$\frac{1 + xyz}{2}, \frac{xy - z}{2}, \frac{yz - x}{2}, \frac{zx - y}{2},$$

resp., to

$$\frac{1 - xyz}{2}, \frac{xy + z}{2}, \frac{yz + x}{2}, \frac{zx + y}{2}$$

The subalgebra $Cl_0(\mathbf{R}^3)$ is identified with the diagonal copy of $\mathbf{H}$ in this decomposition.

(iv) For any inner product space V, we have an isomorphism of algebras $Cl(V) \cong Cl_0(V \oplus \mathbf{R})$. Letting e be a unit vector in $\mathbf{R}$, the isomorphism is given by

$$v_0 + v_1 \mapsto v_0 + v_1 e.$$

It is an easy exercise to show that this map is an isomorphism of algebras. In particular, $Cl_0(\mathbf{R}^4)$ is isomorphic to $\mathbf{H} \oplus \mathbf{H}$.

Comparison with the Exterior Algebra. The grading on $T(V)$ induces an increasing filtration

$$0 \subset \mathcal{F}_1 \subset \mathcal{F}_2 \subset \cdots$$

of $Cl(V)$ by linear subspaces; namely, we set $\mathcal{F}_t$ equal to the image in $Cl(V)$ of

$$\oplus_{n \leq t} \underbrace{V \otimes \cdots \otimes V}_{n \text{ times}}.$$

Clearly, the multiplication in $Cl(V)$ preserves this filtration in the sense that it induces maps

$$\mathcal{F}_i \otimes \mathcal{F}_j \to \mathcal{F}_{i+j}.$$

Thus, there is an associated graded algebra

$$\text{Gr}_{\mathcal{F}_\bullet}(Cl(V)) = \oplus_{n=0}^\infty \mathcal{F}_n/\mathcal{F}_{n-1}$$

with the induced multiplication.

Claim 2.1.1 $\text{Gr}_{\mathcal{F}_\bullet}(Cl(V))$ *is naturally isomorphic to the exterior algebra* $\Lambda^*(V)$.

Proof. Let $\widetilde{\mathcal{F}}_\bullet$ be the increasing filtration of $T(V)$ coming from the grading. Clearly, there is a natural algebra isomorphism between $T(V)$ and $\text{Gr}_{\widetilde{\mathcal{F}}_\bullet}(T(V))$. Consequently, we have a commutative diagram

$$
\begin{array}{ccccccc}
\mathcal{I}\left(v \otimes v + \|v\|^2 1\right) & \longrightarrow & T(V) & \longrightarrow & Cl(V) & \longrightarrow & 0 \\
\downarrow & & \downarrow \simeq & & \downarrow & & \\
\mathcal{I}(v \otimes v) & \longrightarrow & \text{Gr}_{\widetilde{\mathcal{F}}_\bullet}(T(V)) & \longrightarrow & \text{Gr}_{\mathcal{F}_\bullet}(Cl(V)) & \longrightarrow & 0.
\end{array}
$$

Clearly, the quotient of $T(V)$ by the two-sided ideal generated by $v \otimes v$ for $v \in V$ is exactly the exterior algebra $\Lambda^*(V)$. Thus, this diagram induces the claimed isomorphism. $\qquad\square$

There is a natural splitting σ of the map $Cl(V) \to \Lambda^*(V)$. The map σ is linear but not multiplicative. It is defined as follows. Consider an elementary element in $\Lambda^k V$ (; i.e., one contained in $\Lambda^k(W) \subset \Lambda^k(V)$ for some k-dimensional subspace $W \subset V$. Such an element can be written as $re_1 \wedge \cdots \wedge e_k$ where the e_i form an orthonormal basis for W and where $r > 0$. We define

$$\sigma(re_1 \wedge \cdots \wedge e_k) = re_1 \cdots e_k.$$

It is an easy exercise to show that this determines a well-defined mapping which splits the natural projection.

Using this isomorphism, we can view $Cl(V)$ as being given by a new multiplication on $\Lambda^*(V)$. This new multiplication is generated by

$$v \cdot (v_1 \wedge \cdots \wedge v_k) = v \wedge v_1 \wedge \cdots \wedge v_k - v \angle (v_1 \wedge \cdots \wedge v_k)$$

where $\angle$ is the contraction:

$$v \angle (v_1 \wedge \cdots \wedge v_k) = \sum_{i=1}^{k} (-1)^{i-1} \langle v, v_i \rangle v_1 \wedge \cdots \wedge \widehat{v_i} \wedge \cdots \wedge v_k.$$

2.2 The groups $Pin(V)$ and $Spin(V)$

Let $Cl^{\times}(V)$ denote the multiplicative group of units of the algebra $Cl(V)$. We define the group $Pin(V)$ as the subgroup of $Cl^{\times}(V)$ generated by elements $v \in V$ with $\|v\|^2 = 1$. Notice that the given generators of $Pin(V)$ are units since the square of any one of them is -1. We let $Spin(V)$ be the intersection of $Pin(V)$ with $Cl_0(V)$, i.e., the kernel of the group homomorphism $Pin(V) \rightarrow \mathbf{Z}/2\mathbf{Z}$ induced by the splitting of $Cl(V) = Cl_0(V) \oplus Cl_1(V)$. Since the generators of $Pin(V)$ are contained in $Cl_1(V)$, $Spin(V)$ is the subgroup of index two consisting of all elements in $Pin(V)$ which can be written as a product on an even number of the given generators for $Pin(V)$. Let us compute these groups in the first three examples.

(i) The group $Pin(1)$ is the subgroup of $\mathbf{C}$ generated by $\pm i$. Hence it is a cyclic group of order 4. The subgroup $Spin(1)$ is the group of order two ± 1 inside $\mathbf{R}$.

(ii) The group $Pin(2)$ is the subgroup of $\mathbf{H}$ generated by the circle through j and k, i.e., by all elements of the form $\cos(\theta)j + \sin(\theta)k$ for $\theta \in S^1$. It is easy to see that this group is the union of two circles – the usual unit circle in the complex plane and j times it. Hence, the group $Spin(2)$ is isomorphic to S^1.

(iii) The group $Spin(3)$ is isomorphic to the group of unit quaternions in $Cl_0(\mathbf{R}^3) \cong \mathbf{H}$. To see this, notice that under the identification of $Cl(\mathbf{R}^3)$ with $\mathbf{H} \oplus \mathbf{H}$ given above, the vector space $\mathbf{R}^3$ is identified with all pairs $(\alpha, -\alpha)$ where α is a purely imaginary quaternion. It follows that for any pair α, β of purely imaginary unit quaternions the product

$$\alpha\beta \in Spin(V) \subset Cl_0(\mathbf{R}^3) = \mathbf{H} \overset{\Delta}{\hookrightarrow} \mathbf{H} \oplus \mathbf{H}.$$

It is easy to see that this set of products generates the group S^3 of all unit quaternions.

(iv) Let us consider $Spin(4) \subset Cl_0(\mathbf{R}^4) \cong \mathbf{H} \oplus \mathbf{H}$. Of course, under the identification $Cl(\mathbf{R}^3) \cong Cl_0(\mathbf{R}^4)$, the group $Spin(3)$ becomes a subgroup of $Spin(4)$. But there are many different three-dimensional subspaces of $\mathbf{R}^4$. For each such subspace we obtain an embedding of $Spin(3)$ into $Spin(4)$. It is an easy exercise to show that the union of these images generates all of $S^3 \times S^3 \subset \mathbf{H} \times \mathbf{H}$. This then identifies $Spin(4)$ with $SU(2) \times SU(2)$.

Notice that if $\{e_1, \ldots, e_n\}$ is an orthonormal basis for V, then every product $e_{i_1} \cdots e_{i_t}$ is an element of $Pin(V)$. This means that $Pin(V)$ contains a vector space basis for $Cl(V)$, and consequently that $Cl(V)$ is the smallest algebra over $\mathbf{R}$ containing $Pin(V)$ as a subgroup of its multiplicative group of units. Similarly, $Spin(V)$ contains an $\mathbf{R}$-basis for $Cl_0(V)$.

Corollary 2.2.1 • *Two (real or complex) representations of the algebra $Cl_0(V)$ whose restrictions to $Spin(V)$ are isomorphic representations are in fact isomorphic representations of the algebra.*

• *Let A be a (real or complex) module over $Cl_0(V)$ and let $A' \subset A$ be a subspace invariant under the induced action of $Spin(V)$. Then A' is a submodule for the $Cl_0(V)$ action.*

Proof.   Suppose that two modules A and A' for $Cl_0(V)$ admit a linear isomorphism φ which commutes with the induced $Spin(V)$ actions. Then φ commutes with the actions of an $\mathbf{R}$-basis of $Cl_0(V)$ and hence commutes with the $Cl_0(V)$ actions. That is to say, φ is an isomorphism of $Cl_0(V)$-modules. This proves the first result. The second is established similarly. □

Notice that there are analogous results for $Cl(V)$ and $Pin(V)$.

Clearly, the natural action of the group $O(V)$ on V extends to an action of $O(V)$ on $Cl(V)$ as algebra automorphisms preserving the $\mathbf{Z}/2\mathbf{Z}$-grading. This action is effective and hence induces an embedding of $O(V)$ into the algebra automorphisms of $Cl(V)$. Since $Cl(V)$ is generated as an algebra by $V \subset Cl(V)$ and since $v \cdot v = -\|v\|^2 1$, it is easy to see that the image of this embedding consists of all the algebra automorphisms of $Cl(V)$ which preserve the subspace V. The subgroup $SO(V)$ is represented as the group of all algebra automorphisms of $Cl(V)$ which preserve V and act in an orientation-preserving fashion on it.

The group $Spin(V)$ acts on $Cl(V)$ via conjugation: $\sigma \cdot c = \sigma c \sigma^{-1}$. It is easy to see that this action also preserves the algebra structure and the

$\mathbb{Z}/2\mathbb{Z}$-grading. The basic connection between $SO(V)$ and $Spin(V)$ can be seen through these actions.

Lemma 2.2.2 *The conjugation action of $Spin(V)$ on $Cl(V)$ induces a representation of $Spin(V)$ as automorphisms of the Clifford algebra $Cl(V)$. The image of this representation consists of automorphisms which preserve $V \subset Cl(V)$ and which act in an orientation-preserving fashion on V. Thus, we have an induced map $Spin(V) \to SO(V)$. This map is surjective and the kernel is ± 1. If the dimension of V is at least 3, then the kernel of this mapping is the center of $Spin(V)$ and the map presents $Spin(V)$ as the universal covering group of $SO(V)$.*

Proof. First, notice that the representation of $Spin(V)$ on $Cl(V)$ is the restriction of the conjugation representation of $Pin(V)$ on $Cl(V)$. We check that this representation preserves $V \subset Cl(V)$. To do this it suffices to prove the result for the generators of $Pin(V)$, namely the elments of unit length in V. For any $w \in V$ with $\|w\|^2 = 1$ and any $v \in V$, we have

$$wvw^{-1} = -R_{w^\perp}(v).$$

In particular, $wvw^{-1} \in V$. It follows that the action of $Spin(V)$ leaves V invariant and acts in an orientation-preserving fashion on it. Notice that the action of $Spin(V)$ on V consists of all even products of reflections in vectors of length 1. As is well-known, every element of $SO(V)$ is a product of an even number of such reflections. From this it follows easily that the map $Spin(V) \to SO(V)$ is surjective. Clearly, the intersection of $Spin(V)$ with the center of $Cl(V)$ is the kernel of this representation.

Claim 2.2.3 *If the dimension of V is even then the center of the Clifford algebra is $\mathbb{R}$ and is contained in $Cl_0(V)$. If the dimension of V is odd, then the center of V is isomorphic to $\mathbb{R} \oplus \mathbb{R}$ and the intersection of this center with $Cl_0(V)$ is isomorphic to $\mathbb{R}$.*

Proof. Fix an orthonormal basis $\{e_1, \ldots, e_n\}$ for V. A direct computation shows that

$$e_j \cdot (e_{i_1} \cdots e_{i_t}) = \pm (e_{i_1} \cdots e_{i_t}) \cdot e_j$$

where the sign is $(-1)^t$ if $j \neq i_r$ for all $1 \leq r \leq t$ and the sign is $(-1)^{t-1}$ if $j = i_r$ for some $1 \leq r \leq t$. From this it follows easily from the uniqueness of the representatives that when the dimension of V is even the center of $Cl(V)$ is simply the real multiples of the unit 1. When the dimension of V is odd, then the center of $Cl(V)$ is the real vector space spanned by 1 and $e_1 \cdots e_n$. $\square$

It follows that the center of $Spin(V)$ is a subgroup of the multiplicative group $\mathbf{R}^*$. From the fact that the square of any generator of $Pin(V)$ is equal to -1, we see that $\{\pm 1\} \subset Spin(V)$ for all V.

Claim 2.2.4 *The intersection of $Spin(V)$ with the center of $Cl(V)$ is $\{\pm 1\}$.*

Proof. Let us denote by $\phi \mapsto \phi^t$ the antihomomorphism of $Cl(V)$ induced from the map of the tensor algebra that sends $v_1 \otimes \cdots \otimes v_r \mapsto v_r \otimes \cdots \otimes v_1$. It is an easy exercise to show that there is a homomorphism, the norm $N\colon Pin(V) \to \mathbf{R}^*$, defined by $\alpha \to \alpha\epsilon(\alpha^t)$ where if $x = x_0 + x_1 \in Cl(V)$, then $\epsilon(x)$ we mean $x_0 - x_1$. Since $Pin(V)$ is generated by elements of square -1, the map $N\colon Pin(V) \to \mathbf{R}^*$ has image contained in ± 1. In fact, a direct computation shows that $N(v) = 1$ for any $v \in V$ of norm 1. Thus, the image of $N|_{Pin(V)}$ is 1. It is also clear that when restricted to the center of $Cl(V)$, the norm map is the square. This implies that the center of $Spin(V)$ contained in $\{\pm 1\}$, and hence is equal to $\{\pm 1\}$.   □

We have established a natural isomorphism $Spin(V)/\{\pm 1\} \to SO(V)$. Let us show that $Spin(V)$ is not merely two copies of $SO(V)$.

Claim 2.2.5 *Provided that the dimension of V is at least two, the natural map $Spin(V) \to SO(V)$ is a non-trivial double covering*

Proof. We have now shown that $Spin(V) \to SO(V)$ is a double covering. The last thing to establish is that it is a non-trivial double overing if the dimension of V is at least 2. To do this it suffices to restrict to a two-dimensional subspace W of V. Clearly, the preimage of $SO(W) \subset SO(V)$ is $Spin(W) \subset Spin(V)$. Since $\pi_1(SO(W)) \to \pi_1(SO(V))$ is surjective, it suffices to show that $Spin(W) \to SO(W)$ is a non-trivial double covering. We have identified $Cl(W)$ with $\mathbf{H}$ and $Spin(W)$ with the unit circle in $\mathbf{C} \subset \mathbf{H}$. Also, $W \subset Cl(W)$ is identified with the linear subspace generated j, k. Thus, the conjugation action of $Spin(W)$ on W is simple the square of the usual action of S^1 on W. The result follows.   □

This completes the proof of the lemma.   □

Corollary 2.2.6 *Let the dimension of V be n. Then $Spin(V)$ is a compact Lie group of dimension $n(n-1)/2$ which is connected if $n > 1$ and simply connected if $n > 2$. Its Lie algebra is the same as the Lie algebra of $SO(n)$.*

Corollary 2.2.7 *Any (real or complex) module for $Cl_0(V)$ or $Cl(V)$ is completely reducible; i.e., can be decomposed into a direct sum of irreducible modules.*

Proof. Since $Spin(V)$ and $Pin(V)$ are compact groups, the corresponding statements for representations of these groups holds. From this and Corollary 2.2.1, the result follows immediately. □

Example. We have identified $Cl(\mathbf{R}^3)$ with the algebra $\mathbf{H} \oplus \mathbf{H}$ and $Cl_0(\mathbf{R}^3)$ as the diagonal copy of $\mathbf{H}$. The group $Spin(3)$ is a subgroup of units of norm one in $Cl_0(\mathbf{R}^3) = \mathbf{H}$. Since $Spin(3)$ has dimension three, it follows that $Spin(3)$ is the group of units of norm one in $\mathbf{H}$. This group is identified with S^3 or with $SU(2)$. The action of $Spin(3)$ on $\mathbf{R}^3$ is given by restricting the usual conjugation action of S^3 on the three-dimensional space of purely imaginary quaternions. This of course is the usual adjoint action of $SU(2)$ on its Lie algebra, and the image of this representation is $SO(3)$. The Clifford algebras have allowed us to generalize this construction to higher dimensions even though there are not division algebras generalizing $\mathbf{H}$ in higher dimensions. What we have done in higher dimensions is to replace $\mathbf{H}$ by a Clifford algebra instead of a division algebra.

(ii) The group $Spin(4)$ is the double covering of $SO(4) \cong SU(2) \times SU(2)/\{\pm 1\}$. Thus, $Spin(4) \cong SU(2) \times SU(2)$. We will see more later about this decomposition.

2.3 Splitting of the Clifford Algebra

Let V be an oriented real inner product space (with a positive definite inner product). Let $Cl(V) \otimes_{\mathbf{R}} \mathbf{C}$ be the complexification of the Clifford algebra. It is a complex algebra. Fix an oriented orthonormal basis $\{e_1, \ldots, e_n\}$ for V. We define

$$\omega_{\mathbf{C}} = i^{\left[\frac{n+1}{2}\right]} e_1 \cdots e_n.$$

A simple computation shows that $\omega_{\mathbf{C}}^2 = 1$ and that $\omega_{\mathbf{C}}$ is independent of the choice of oriented orthonormal basis. Thus, we can use $\omega_{\mathbf{C}}$ to give a canonical decomposition of $Cl(V) \otimes \mathbf{C}$ into $(Cl(V) \otimes \mathbf{C})^+ \oplus (Cl(V) \otimes \mathbf{C})^-$ where $(Cl(V) \otimes \mathbf{C})^{\pm}$ is the eigenspace of ± 1 for the action of $\omega_{\mathbf{C}}$ by left multiplication. Notice that if the dimension of V is odd, then $\omega_{\mathbf{C}}$ is in the center of $Cl(V) \otimes \mathbf{C}$ whereas if the dimension of V is even then $\omega_{\mathbf{C}}$ is in the center of $Cl_0(V) \otimes \mathbf{C}$ but anti-commutes with elements in $Cl_1(V) \otimes \mathbf{C}$.

If $\dim(V) = n$ is odd, then $\omega_{\mathbf{C}}$ is contained in the center of $Cl(V) \otimes \mathbf{C}$. This means that $(Cl(V) \otimes \mathbf{C})^{\pm}$ are subalgebras which annihilate each other. That is to say we have an orthogonal decomposition of algebras

$$Cl(V) \otimes \mathbf{C} = (Cl(V) \otimes \mathbf{C})^{+} \oplus (Cl(V) \otimes \mathbf{C})^{-}.$$

Lemma 2.3.1 *If the dimension of V is odd, then the algebras* $(Cl(V) \otimes \mathbf{C})^{\pm}$ *are both isomorphic to* $Cl_0(V) \otimes \mathbf{C}$.

Proof. Since multiplication by $\omega_{\mathbf{C}}$ interchanges $Cl_0(V) \otimes \mathbf{C}$ and $Cl_1(V) \otimes \mathbf{C}$, it follows that $Cl_0(V) \otimes \mathbf{C}$ meets both $(Cl(V) \otimes \mathbf{C})^{\pm}$ trivially. Thus, the compositions

$$Cl_0(V) \otimes \mathbf{C} \hookrightarrow Cl(V) \otimes \mathbf{C} \xrightarrow{\pi^{\pm}} (Cl(V) \otimes \mathbf{C})^{\pm}$$

are isomorphisms of algebras.   □

Notice that this embedding of $Cl_0(V) \otimes \mathbf{C}$ is the graph of an algebra isomorphism from $(Cl(V) \otimes \mathbf{C})^{+} \to (Cl(V) \otimes \mathbf{C})^{-}$.

Examples.   (i) Since $Cl(\mathbf{R}^1) \cong \mathbf{C}$, we see that $Cl(\mathbf{R}^1) \otimes \mathbf{C}$ is isomorphic to $\mathbf{C} \otimes_{\mathbf{R}} \mathbf{C}$. The decomposition of $Cl(\mathbf{R}^1) \otimes \mathbf{C}$ into $(Cl(\mathbf{R}^1) \otimes \mathbf{C})^{\pm}$ corresponds to the usual decomposition

$$\mathbf{C} \otimes_{\mathbf{R}} \mathbf{C} \cong \mathbf{C} \oplus \mathbf{C}.$$

Thus, in this case the decomposition of the complexification does not come from a decomposition of the real Clifford algebra, which is a simple algebra.

(ii) We have an identification of $Cl(\mathbf{R}^3)$ with $\mathbf{H} \oplus \mathbf{H}$. The complexification of this splitting is the splitting of $Cl(\mathbf{R}^3) \otimes \mathbf{C}$ into $(Cl(\mathbf{R}^3) \otimes \mathbf{C})^{\pm}$. Thus, in this case the splitting of the complex algebra is induced from a splitting of the real algebra.

Lemma 2.3.2 *If the dimension of V is congruent to 3 modulo 4, then* $Cl(V)$ *splits as an orthogonal sum of two algebras* $Cl(V) = Cl(V)^{+} \oplus Cl(V)^{-}$ *induces the above splitting on the complexified algebras.*

Proof. When the dimension of V is congruent to 3 modulo four then the complex unit $\omega_{\mathbf{C}}$ is given by $(-1)^{\frac{n+1}{4}} e_1 \cdots e_n$ and hence is contained in the real algebra. Thus, its ± 1 eigenspaces are real subspaces.   □

In the case when the dimension of V is even, $\omega_{\mathbf{C}}$ produces a different type of decomposition of $Cl(V) \otimes \mathbf{C}$. In this case $\omega_{\mathbf{C}}$ is in the center of $Cl_0(V) \otimes \mathbf{C}$ and hence induces a splitting $Cl_0(V) \otimes \mathbf{C} = (Cl_0(V) \otimes \mathbf{C})^+ \oplus (Cl_0(V) \otimes \mathbf{C})^-$. Under the isomorphism of $Cl_0(V)$ with $Cl(W)$ for a codimension-1 subspace $W \subset V$, this decomposition agrees with the decomposition given before in the odd dimensional case. In particular, there is an isomorphism of algebras between $(Cl_0(V) \otimes \mathbf{C})^+$ and $(Cl_0(V) \otimes \mathbf{C})^-$.

Lemma 2.3.3 *If V is even dimensional, then $(Cl_0(V) \otimes \mathbf{C})^+$ is isomorphic as an algebra to $Cl(W) \otimes \mathbf{C}$ where $W \subset V$ is a codimension-2 subspace.*

Proof. We have an isomorphism $Cl_0(V) \otimes \mathbf{C}$ with $Cl(W') \otimes \mathbf{C}$ where $W' \subset V$ is a codimension-1 subspace. Thus,

$$(Cl_0(V) \otimes \mathbf{C})^+ \cong (Cl(W') \otimes \mathbf{C})^+ \cong Cl_0(W') \otimes \mathbf{C}.$$

The latter is isomorphic to $Cl(W) \otimes \mathbf{C}$ for a codimension-1 subspace $W \subset W'$. □

Similarly, there is a decomposition

$$Cl_1(V) \otimes \mathbf{C} = (Cl_1(V) \otimes \mathbf{C})^+ \oplus (Cl_1(V) \otimes \mathbf{C})^-.$$

While we shall not prove this now, this decomposition into four pieces corresponds to the fact that $Cl(V) \otimes \mathbf{C}$ is a two-by-two matrix algebra over $(Cl_0(V) \otimes \mathbf{C})^+$.

Comparison of $(Cl(V) \otimes \mathbf{C})^+$ and $\Lambda_+^*(V) \otimes \mathbf{C}$. The main result that we wish to establish here is the following:

Lemma 2.3.4 *Suppose that V is a vector space of dimension 4. Under the natural vector space isomorphism $\Lambda^*(V) \otimes \mathbf{C} \cong Cl(V) \otimes \mathbf{C}$, the subspace corresponding to $(Cl_0(V) \otimes \mathbf{C})^+$ is $\mathbf{C}(\frac{1+\omega}{2}\mathbf{C}) \oplus (\Lambda_+^2(V) \otimes \mathbf{C})$.*

Proof. Of course, $Cl_0(V)$ corresponds to $\Lambda^0(V) \oplus \Lambda^2(V) \oplus \Lambda^4(V)$. Multiplication by $\omega_{\mathbf{C}}$ leaves $\Lambda^2(V) \otimes \mathbf{C}$ invariant and switches $\Lambda^0(V) \otimes \mathbf{C}$ and $\Lambda^4(V) \otimes \mathbf{C}$. Fix an oriented orthonormal basis $\{e_1, \ldots, e_4\}$ for V. Then

$$\omega_{\mathbf{C}} \cdot (e_1 e_2) = e_3 e_4.$$

By symmetry it follows that $\omega_{\mathbf{C}} e_{i_1} e_{i_2} = *(e_{i_1} e_{i_2})$ for any distinct indices i_1, i_2. The result in now immediate. (Notice that this result would not hold if we used the real unit $\omega_{\mathbf{R}} = e_1 e_2 e_3 e_4$ instead of the complex unit.) □

2.4   The complexification of the $Cl(V)$

Let begin with the first two examples.

(i) We have seen that $Cl(\mathbf{R}^1) \cong \mathbf{C}$ and hence $Cl(\mathbf{R}^1) \otimes \mathbf{C} \cong \mathbf{C} \oplus \mathbf{C}$.

(ii) We have seen that $Cl(\mathbf{R}^2) \cong \mathbf{H}$ and hence that $Cl(\mathbf{R}^2) \otimes \mathbf{C} \cong \mathbf{H} \otimes \mathbf{C}$. We define a map from $\mathbf{H} \to \mathbf{C}[2]$, the algebra of two-by-two complex matrices, by sending $\alpha + j\beta$ to

$$\begin{pmatrix} \alpha & -\overline{\beta} \\ \beta & \overline{\alpha} \end{pmatrix}.$$

Writing elements of $\mathbf{H}$ as $x + jy$ for $x, y \in \mathbf{C}$, this matrix is gives the action of $\alpha + j\beta$ by left multiplication on $\mathbf{H}$ viewed as a two-dimensional vector space over $\mathbf{C}$. Extending scalars we get a homomorphism of $\mathbf{C}$-algebras

$$\mathbf{H} \otimes \mathbf{C} \to \mathbf{C}[2].$$

Clearly, this extension is an isomorphism of $\mathbf{C}$-algebras. This gives us an identification of $Cl(\mathbf{R}^2) \otimes \mathbf{C}$ with the matrix algebra $\mathbf{C}[2]$ of two-by-two complex matrices. Notice that $Cl(\mathbf{R}^2) \otimes \mathbf{C}$ is a two-by-two matrix algebra over $\mathbf{C} \cong (Cl_0(\mathbf{R}^2) \otimes \mathbf{C})^+$.

It turns out that these two computations are enough to determine the structure of the complexifications of all the Clifford algebras by induction. The basic inductive step is contained in the next result.

Lemma 2.4.1

$$Cl(V \oplus \mathbf{R}^2) \otimes \mathbf{C} \cong (Cl(V) \otimes_{\mathbf{R}} \mathbf{C}) \otimes_{\mathbf{C}} (Cl(\mathbf{R}^2) \otimes_{\mathbf{R}} \mathbf{C}).$$

Proof.   Let $v_1, \ldots, v_n$ be an orthonormal basis for V and let e_1, e_2 be the standard basis for $\mathbf{R}^2$. We define a real linear map

$$V \oplus \mathbf{R}^2 \to (Cl(V) \otimes_{\mathbf{R}} \mathbf{C}) \otimes_{\mathbf{C}} (Cl(\mathbf{R}^2) \otimes \mathbf{C})$$

by sending v_j to $iv_j \otimes e_1 e_2$ for all $1 \leq j \leq n$ and by sending e_r to $1 \otimes e_r$. A direct check shows that this map satisfies the condition to extend to an algebra homomorphism

$$Cl(V) \to (Cl(V) \otimes_{\mathbf{R}} \mathbf{C}) \otimes_{\mathbf{C}} (Cl(\mathbf{R}^2) \otimes \mathbf{C})$$

and the by extension of scalars to a map from $Cl(V) \otimes_{\mathbf{R}} \mathbf{C}$ to this tensor product.

Clearly, the domain and range have the same dimension and the map is onto algebra generators for the range. Hence this map is an isomorphism of algebras. □

Corollary 2.4.2 *If the dimension of V is $2n$, then $Cl(V) \otimes_{\mathbf{R}} \mathbf{C}$ is isomorphic to the matrix algebra $\mathbf{C}[2^n]$. If the dimension of V is $2n+1$, then $Cl(V) \otimes_{\mathbf{R}} \mathbf{C}$ is isomorphic as an algebra to the direct sum of two copies of $\mathbf{C}[2^n]$.*

Proof. This is easily proved by induction from the previous lemma, the computations for V of dimension one and two, and the elementary fact that $\mathbf{C}[m] \otimes_{\mathbf{C}} \mathbf{C}[k]$ is isomorphic to $\mathbf{C}[km]$. $\qquad\qquad\qquad\qquad\qquad$ $\square$

Corollary 2.4.3 *If the dimension of V is $2n$, then $(Cl_0(V) \otimes \mathbf{C})^+$ is isomorphic to the matrix algebra $\mathbf{C}[2^{n-1}]$.*

Proof. This is immediate from the previous corollary and Lemma 2.3.3. $\qquad\qquad\qquad\qquad\qquad\qquad\qquad\qquad\qquad\qquad\qquad$ $\square$

As we remarked above, when the dimension of V is odd, then the complex unit $\omega_{\mathbf{C}}$ is contained in the center of $Cl(V) \otimes \mathbf{C}$. The decomposition of $Cl(V) \otimes \mathbf{C}$ into a direct sum of matrix algebras is the decomposition into the ± 1 eigenspaces of $\omega_{\mathbf{C}}$. Thus, the map $\alpha \mapsto \epsilon(\alpha)$ switches the two summands, inducing an isomorphism between them. It follows that $Cl_0(V) \otimes \mathbf{C}$ is embedded in a diagonal fashion with respect to this isomorphism.

Corollary 2.4.4 *If the dimension of V is $2n$, then $Cl(V)$ has a unique irreducible, finite dimensional, complex representation $S_{\mathbf{C}}(V)$ up to isomorphism. Any such representation has dimension 2^n. The action of $Cl(V) \otimes \mathbf{C}$ on $S_{\mathbf{C}}(V)$ induces an isomorphism*

$$Cl(V) \otimes \mathbf{C} \to \text{End}_{\mathbf{C}}(S_{\mathbf{C}}(V)) = S_{\mathbf{C}}(V) \otimes S_{\mathbf{C}}(V)^*.$$

If the dimension of V is $2n + 1$, then $Cl(V)$ has exactly two irreducible, finite dimensional, complex representations up to isomorphism. These induce isomorphic representations $S_{\mathbf{C}}(V)$ of $Cl_0(V)$ by restriction. Any such representation has dimension 2^n, and the action of Clifford multiplication induces a map

$$Cl_0(V) \otimes \mathbf{C} \to \text{End}(S_{\mathbf{C}}(V)) = S_{\mathbf{C}}(V) \otimes S_{\mathbf{C}}(V)^*$$

which is an isomorphism.

Proof. Wederburn's theorem tells us that $\mathbf{C}[n]$ has a unique irreducible, finite dimensional, complex representation $S^n_{\mathbf{C}}$ up to isomorphism. Furthermore, for this representation the map

$$\mathbf{C}[n] \to \mathrm{End}(S^n_{\mathbf{C}})$$

is an isomorphism of algebras. The result in the case when the dimension of V is even is immediate from this. In the case the dimension of V is odd, it follows from Wederburn's theorem that there are exactly two non-isomorphic irreducible, finite dimensional, complex representations of $Cl(V)$. They are obtained from projecting $Cl(V) \otimes \mathbf{C}$ onto one of the two factors and the taking the irreducible representation of that factor. Since $Cl_0(V) \otimes \mathbf{C}$ is embedded diagonally in the direct sum of the matrix algebras, it is clear that the restriction of the non-isomorphic irreducible complex representations of $Cl(V)$ to $Cl_0(V)$ become isomorphic. Since these representations are irreducible, the last statement also follows from Wederburn's theorem. □

Corollary 2.4.5 *Let us suppose that the dimension of V is even. Let $S_{\mathbf{C}}(V)$ be an irreducible (complex) representation of $Cl(V) \otimes \mathbf{C}$. Then $S_{\mathbf{C}}(V)$ decomposes into $S^{\pm}_{\mathbf{C}}(V)$ under the action of $\omega_{\mathbf{C}}$. This decomposition is a decomposition of modules over $Cl_0(V) \otimes \mathbf{C}$; whereas the action of $Cl_1(V) \otimes \mathbf{C}$ interchanges $S^{\pm}_{\mathbf{C}}(V)$. Clifford multiplication induces isomorphisms*

$$(Cl_0(V) \otimes \mathbf{C})^+ \cong \mathrm{End}_{\mathbf{C}}(S^+_{\mathbf{C}}(V))$$

$$(Cl_0(V) \otimes \mathbf{C})^- \cong \mathrm{End}_{\mathbf{C}}(S^-_{\mathbf{C}}(V))$$

$$(Cl_1(V) \otimes \mathbf{C})^- \cong \mathrm{Hom}_{\mathbf{C}}(S^+_{\mathbf{C}}(V), S^-_{\mathbf{C}}(V))$$

$$(Cl_1(V) \otimes \mathbf{C})^+ \cong \mathrm{Hom}_{\mathbf{C}}(S^-_{\mathbf{C}}(V), S^+_{\mathbf{C}}(V)).$$

Consequently, $S^{\pm}_{\mathbf{C}}(V)$ are the two inequivalent irreducible representations of $Cl_0(V) \otimes \mathbf{C}$ up to isomorphism.

Proof. It is easy to establish that the action of the various pieces of $Cl(V) \otimes \mathbf{C}$ is as claimed. Since Clifford multiplication induces an isomorphism

$$Cl(V) \cong \mathrm{End}(S_{\mathbf{C}}(V)),$$

the result then follows easily. □

Corollary 2.4.6 *There is a unique complex representation of $Spin(V)$ (up to isomorphism) induced from any irreducible complex finite dimensional representation of $Cl(V)$. This representation is called the complex spin representation and is denoted*

$$\Delta_{\mathbf{C}} \colon Spin(V) \to \mathrm{Aut}_{\mathbf{C}}(S_{\mathbf{C}}(V)).$$

Proof. This is immediate from the previous result and the fact that by definition $Spin(V)$ is contained in $Cl_0(V)$. $\qquad\square$

2.5 The Complex Spin Representation

Let $\Delta_{\mathbf{C}} \colon Spin(V) \times S_{\mathbf{C}}(V) \to S_{\mathbf{C}}(V)$ be the complex spin representation.

Proposition 2.5.1 *If the dimension of V is even, say $2n$, then this representation decomposes into two inequivalent irreducible representations $S_{\mathbf{C}}^{+}(V)$ and $S_{\mathbf{C}}^{-}(V)$ of $Spin(V)$, each of dimension 2^{n-1}. These representations of $Spin(V)$ are denoted by $\Delta_{\mathbf{C}}^{\pm}$. If the dimension of V is odd, say $2n+1$, then $\Delta_{\mathbf{C}}$ is an irreducible representation of $Spin(V)$ of dimension 2^{n}.*

Proof. This is immediate from the previous Corollaries 2.2.1 and 2.4.4. $\qquad\square$

Notice that we are not claiming (and indeed, it is not true) that the only irreducible complex representation of $Spin(2k+1)$ is $\Delta_{\mathbf{C}}$ and the only irreducible complex representations of $Spin(2k)$ are $\Delta_{\mathbf{C}}^{\pm}$. It is simply that these irreducible representations of $Spin(n)$ are distinguished as being the only ones that extend to (automatically irreducible) representations of $Cl_0(V)$.

Examples. (i) $Cl(\mathbf{R}^2) \cong \mathbf{H}$ and hence $Cl(V) \otimes \mathbf{C}$ is isomorphic to $\mathbf{C}[2]$. Hence the spin representation is

$$\Delta_{\mathbf{C}} \colon Spin(\mathbf{R}^2) \to \mathrm{Aut}(\mathbf{C}^2).$$

This representation decomposes as a sum of two one-dimensional complex representations

$$\Delta_{\mathbf{C}}^{\pm} \colon Spin(\mathbf{R}^2) \to \mathrm{Aut}(S_{\mathbf{C}}^{\pm}(\mathbf{R}^2)).$$

Of course, $Spin(\mathbf{R}^2) \cong S^1$ embedded in the standard way in $\mathbf{C} \subset \mathbf{H}$. As we have seen, under the embedding $\mathbf{H} \subset \mathbf{C}[2]$ this circle is embedded as

$$\alpha \in S^1 \mapsto \begin{pmatrix} \alpha & 0 \\ 0 & \bar{\alpha} \end{pmatrix}.$$

Since $\omega_{\mathbf{C}} = ie_1e_2$, the element e_1e_2 acts by $-i$ on $S_{\mathbf{C}}^+(\mathbf{R}^2)$, and hence $S_{\mathbf{C}}^+(\mathbf{R}^2)$ has the conjugate to the standard action of S^1 on $\mathbf{C}$ and $S_{\mathbf{C}}^-(\mathbf{R}^2)$ has the standard action.

(ii) As we have seen $Spin(\mathbf{R}^3)$ is isomorphic to $SU(2)$. The spin representation $\Delta_{\mathbf{C}}$ in this case is the standard representation of $SU(2)$ on $\mathbf{C}^2$.

(iii) As we have seen $Spin(\mathbf{R}^4)$ is isomorphic to $SU(2) \times SU(2)$. The spin representation $\Delta_{\mathbf{C}}^+$ is simply the projection of $Spin(\mathbf{R}^4)$ onto the first factor followed by the standard representation of $SU(2)$ on $\mathbf{C}^2$. Similarly, $\Delta_{\mathbf{C}}^-$ is given by the composition of the projection onto the second factor followed by the standard representation. Clearly, these representation of $Spin(\mathbf{R}^4)$ are inequivalent but they are equivalent under an outer automorphism of $Spin(\mathbf{R}^4)$.

2.6   The Group $Spin^c(V)$

By definition $Spin^c(V)$ is the subgroup of the multiplicative group of units of $Cl(V) \otimes_{\mathbf{R}} \mathbf{C}$ generated by $Spin(V)$ and the unit circle of complex scalars.

Lemma 2.6.1 *There is an isomorphism $Spin^c(V) \cong Spin(V) \times_{\{\pm 1\}} S^1$.*

Proof.   Since the circle of unit scalars is in the center of $Cl(V) \otimes \mathbf{C}$ it commutes with $Spin(V)$. Thus, from the definition we have a natural surjective map

$$Spin(V) \times S^1 \to Spin^c(V).$$

The kernel of this mapping is simply the pairs (α, α^{-1}) where $\alpha \in Spin(V) \cap S^1$. But we have already seen that the intersection of $Spin(V)$ with the scalars is ± 1. The result follows. $\Box$

If we divide $Spin^c(V)$ by the subgroup $\{\pm 1\}$, the quotient is $SO(V) \times S^1$. Thus, $Spin^c(V)$ is the double covering group of $SO(V) \times S^1$ which is non-trivial on each factor. This is precisely the same as the pre-image under the map $Spin(V \oplus \mathbf{R}^2) \to SO(V \oplus \mathbf{R}^2)$ of the natural embedding $SO(V) \times S^1 \subset SO(V \oplus \mathbf{R}^2)$. The conjugation action of $Spin(V)$ on $Cl(V)$

extends to the conjugation action of $Spin^c(V)$ on $Cl(V) \otimes \mathbf{C}$. This later action preserves the real Clifford algebra and has the same image $SO(V)$ as the conjugation action of $Spin(V)$. In particular, the kernel of the action is $\{\pm 1\} \times_{\{\pm 1\}} S^1 \cong S^1$.

Notice that $Spin(V)$ is naturally identified with the subgroup

$$Spin(V) \times_{\{\pm 1\}} \{\pm 1\}$$

of $Spin^c(V)$.

Lemma 2.6.2 *Let* $\rho \colon Spin(V) \to GL_{\mathbf{C}}(W)$ *be a complex representation. Suppose that* $\rho(-1) = -1$. *Then there is a unique extension of* ρ *to a representation* $\hat{\rho} \colon Spin^c(V) \to GL_{\mathbf{C}}(W)$.

Proof. Since ρ is a complex representation, it commutes with the multiplication of the complex scalars of unit length $S^1 \times W \to W$. Thus, ρ extends to a map $\rho' \colon Spin(V) \times S^1 \to GL_{\mathbf{C}}(W)$. The fact that $\rho(-1) = -1$ means that this representation descends to one $\hat{\rho}$ as required. $\quad\square$

Corollary 2.6.3 *Let* $\Delta_{\mathbf{C}} \colon Spin(V) \to GL_{\mathbf{C}}(S_{\mathbf{C}}(V))$ *be the complex spin representation. Then there is a unique extension of this representation to* $\hat{\Delta}_{\mathbf{C}} \colon Spin^c(V) \to GL_{\mathbf{C}}(S_{\mathbf{C}}(V))$. *Furthermore, if the dimension of* V *is even, then* $\hat{\Delta}_{\mathbf{C}}$ *splits as* $\hat{\Delta}_{\mathbf{C}}^+ + \hat{\Delta}_{\mathbf{C}}^-$ *where* $\hat{\Delta}_{\mathbf{C}}^{\pm}$ *is the unique extension of* $\Delta_{\mathbf{C}}^{\pm}$.

Proof. The only thing to show is that $\Delta_{\mathbf{C}}(-1) = -1$. But this is clear since $\Delta_{\mathbf{C}}$ is the restriction of a representation of the $\mathbf{C}$-algebra $Cl(V) \otimes_{\mathbf{R}} \mathbf{C}$. $\quad\square$

Chapter 3

Spin Bundles and the Dirac Operator

3.1 Spin Bundles and Clifford Bundles

Spin Bundles. Fix a real vector space V with a positive definite inner product. We suppose throughout this chapter that the dimension of V is at least 2. Suppose that $P \to X$ is a principal $SO(V)$ bundle. We wish to understand when this bundle lifts to a principal $Spin(V)$ bundle; that is to say, when there is a principal $Spin(V)$ bundle $\widetilde{P} \to X$ whose quotient by the center ± 1 of $Spin(V)$ is isomorphic as an $SO(V)$ bundle to P. This is of course, a standard problem in algebraic topology or obstruction theory whose solution is well-known.

Lemma 3.1.1 *The bundle $P \to X$ lifts to a principal $Spin(V)$ bundle if and only if the second Stiefel-Whitney class $w_2(P) \in H^2(X; \mathbf{Z}/2\mathbf{Z})$ is equal to zero.*

Another way to say this is that $P \to X$ lifts if and only if there is a homomorphism $\pi_1(P) \to \mathbf{Z}/2\mathbf{Z}$ which restricts to the non-trivial homomorphism $\pi_1(SO(V)) \to \mathbf{Z}/2\mathbf{Z}$, or equivalently if the differential $d_2 \colon H_2(X; \mathbf{Z}/2\mathbf{Z}) \to H_1(SO(V); \mathbf{Z}/2\mathbf{Z})$ is trivial. Equivalently, $P \to X$ lifts if and only if P is trivial over the two-skeleton of X. If there is a lifting, then the set of liftings up to isomorphism form a torseur of the group $H^1(X; \mathbf{Z}/2\mathbf{Z})$.

Such a lifting is called a *spin structure for P*. In the special case when P is the tangent (or cotangent) bundle of a riemannian manifold $\widetilde{P}$ is called

23

a *spin structure for the manifold.*

Suppose that $P \to X$ is a principal $SO(n)$-bundle with a spin structure $\tilde{P} \to X$. Then there is an associated complex spin bundle

$$\tilde{P} \times_{Spin(n)} S_{\mathbf{C}}(\mathbf{R}^n)$$

induced by the rerpesentation $\Delta_{\mathbf{C}} \colon Spin(n) \to \operatorname{Aut}(S_{\mathbf{C}}(\mathbf{R}^n))$. We denote this bundle by $S_{\mathbf{C}}(\tilde{P})$. If P is an $SO(n)$-bundle for n odd, then $S_{\mathbf{C}}(\tilde{P})$ is of (complex) dimension $2^{\frac{n-1}{2}}$ If n is even, then the bundle decomposes into a direct sum

$$S_{\mathbf{C}}(\tilde{P}) = S_{\mathbf{C}}^+(\tilde{P}) \oplus S_{\mathbf{C}}^-(\tilde{P})$$

where

$$S_{\mathbf{C}}^{\pm}(\tilde{P}) = \tilde{P} \times_{Spin(n)} S_{\mathbf{C}}^{\pm}(\mathbf{R}^n)$$

corresponding to the decomposition of $\Delta_{\mathbf{C}}$ into $\Delta_{\mathbf{C}}^+ \oplus \Delta_{\mathbf{C}}^-$. These bundles are called the *plus and minus spin bundles* associated with $\tilde{P}$. They are complex vector bundles of complex dimension $2^{\frac{n}{2}-1}$. As we have seen, since $Spin(n)$ is compact, these bundles carry hermitian inner products unique up to isomorphism. In fact since the bundles are induced by an action of the Clifford algebra on $S_{\mathbf{C}}(\mathbf{R}^n)$, we can choose the metric to be invariant under the action of $Pin(\mathbf{R}^n)$. This means that Clifford multiplication by a unit vector in $\mathbf{R}^n \subset Cl(\mathbf{R}^n)$ is an isometry of $S_{\mathbf{C}}(\mathbf{R}^n)$. We always implicitly assume that we have chosen such a metric.

Spin^c-bundles. Let us consider the analogous questions to the ones considered above for the group *Spin^c* instead of *Spin*. We shall consider the map $Spin^c(n) \to SO(n)$ given by dividing out by the center, and ask when a principal $SO(n)$-bundle $P \to X$ lifts to a principal $Spin^c(n)$-bundle. The homomorphism $Spin^c(n) \to S^1$ given by dividing out by $Spin(n)$ determines a complex line bundle $\mathcal{L} \to X$ associated to any principal $Spin^c(n)$-bundle. This is called the *determinant line bundle* of the $Spin^c(n)$ bundle. If this $Spin^c(n)$ bundle lifts a principal $SO(n)$-bundle $P \to X$, then it is easy to see that this determinant line bundle has a first Chern class $c_1(\mathcal{L})$ which agrees modulo two with $w_2(P)$. Conversely, given any line bundle $\mathcal{L} \to X$ whose first Chern class satisfies this mod 2 equation, there is a $Spin^c(n)$ lifting of P with determinant line bundle isomorphic to $\mathcal{L}$. From the exact sequence

$$\{1\} \to S^1 \to Spin^c(n) \to SO(n) \to \{1\}$$

we see that the liftings of P to a principal $Spin^c(n)$ bundle form a torseur of $H^2(X; \mathbf{Z})$. Varying a $Spin^c(n)$-lifting by a class $\alpha \in H^2(X; \mathbf{Z})$ changes the first Chern class of the determinant line bundle by 2α.

Given a $Spin^c(n)$-lifting $\widetilde{P}$ of a principal $SO(n)$-bundle then we can construct the $Spin^c$ bundle

$$S_{\mathbf{C}}(\widetilde{P}) = \widetilde{P} \times_{Spin^c(n)} S_{\mathbf{C}}(\mathbf{R}^n).$$

It is a complex vector bundle. Varying the $Spin^c(n)$-lifting by a class $\alpha \in H^2(X; \mathbf{Z})$ has the effect of replacing $S_{\mathbf{C}}(\widetilde{P})$ by $S_{\mathbf{C}}(\widetilde{P}) \otimes L_\alpha$ where L_α is the complex line bundle with first Chern class α.

$Spin^c(4)$-**liftings for the orthogonal frame bundle of a 4-manifold.** One of the great advantages of $Spin^c$-structures in studying manifolds of dimension four is that every oriented 4-manifold possesses one.

Lemma 3.1.2 *Let X be an oriented four-manifold and let $P \to X$ be the frame bundle of the tangent bundle. Then there is a lifting $\widetilde{P}$ of P to a $Spin^c(4)$-bundle.*

Proof. We need only see that $w_2(X)$ lifts to an integral class $c \in H^2(X; \mathbf{Z})$ in order to prove the existence of a $Spin^c$-lifting. But for any class $x \in H_2(X; \mathbf{Z}/2\mathbf{Z})$ the value of $w_2(X)$ on x is given as follows: Represent x by an embedded (possibly non-orientable) closed surface in X and take the self-intersection of this surface modulo two. To see that $w_2(X)$ lifts to an integral class, we must see that its integral Bockstein $\delta w_2(X)$ is zero. But this torsion integral class is zero if an only if it evaluates trivially on every $\mathbf{Z}/2^k\mathbf{Z}$ class of dimension three. Any such class is represented by a mapping of a smooth $\mathbf{Z}/2^k\mathbf{Z}$-manifold into X. The value of $\delta w_2(X)$ on such a class is equal to the value of $w_2(X)$ on the Bockstein of this $\mathbf{Z}/2^k\mathbf{Z}$-manifold. Thus, we need only see that $w_2(X)$ vanishes on integral classes which represent torsion elements in $H_2(X; \mathbf{Z})$. But this is clear, any such class is represented by a smoothly embedded oriented surface with zero self-intersection. $\qquad\square$

Clifford Bundles and their actions on the Spin bundles. Let P be an $SO(n)$-bundle. Not only can we form the associated complex spin bundles to a $Spin$ or $Spin^c$-bundle $\widetilde{P}$ lifting P, we can also form bundles of complex Clifford algebras associated to P. Notice that since $SO(n)$ acts on the Clifford algebra $Cl(\mathbf{R}^n)$, wye can associate to $P \to X$ a bundle

$$Cl(P) = P \times_{SO(n)} Cl(\mathbf{R}^n).$$

This is a locally trivial bundle of Clifford algebras. Notice that it is defined without the need of a spin structure. We also have the complex version

$$Cl(P) \otimes \mathbf{C} = P \times_{SO(n)} (Cl(\mathbf{R}^n) \otimes \mathbf{C}),$$

a bundle of complexified Clifford algebras. Notice that these bundles of algebras decompose: $Cl(P) = Cl_0(P) \oplus Cl_1(P)$. Also because the element $\omega_{\mathbf{C}}$ is invariant under the action of $SO(V)$, it follows that there $\omega_{\mathbf{C}}$ defines a section of $Cl(P)$ of square 1. Using this section we can produce a decomposition

$$Cl(P) \otimes \mathbf{C} = (Cl(P) \otimes \mathbf{C})^+ \oplus (Cl(P) \otimes \mathbf{C})^-.$$

When P is an $SO(n)$ bundle for n odd, this is a decomposition into orthogonal bundles of algebras.

In the presence of a spin or $Spin^c$ structure $\tilde{P}$ on P these Clifford bundles act on the complex spin bundles. Let $S_{\mathbf{C}}(\tilde{P}) \to X$ be the associated complex spin bundle. We can make $S_{\mathbf{C}}(\tilde{P})$ a bundle of modules over the bundle of complex algebras $Cl(P) \otimes \mathbf{C}$ in the following way. The bundle $Cl(P)$ is also the bundle of algebras associated to $\tilde{P}$ and the conjugation action of $Spin(\mathbf{R}^n)$ or $Spin^c(\mathbf{R}^n)$ on $Cl(\mathbf{R}^n)$; i.e., the action given by

$$\alpha \cdot \lambda = \alpha \lambda \alpha^{-1}.$$

It is easy to see that Clifford multiplication of $Cl(\mathbf{R}^n) \otimes \mathbf{C}$ on $S_{\mathbf{C}}(\mathbf{R}^n)$ commutes with these actions of $Spin(n)$ or $Spin^c(n)$. That is to say: if $\lambda \in Cl(\mathbf{R}^n) \otimes \mathbf{C}$, $\sigma \in S_{\mathbf{C}}(n)$ and $\alpha \in Spin(n)$ or $\alpha \in Spin^c(n)$ then we have

$$(\alpha \lambda \alpha^{-1}) \cdot (\alpha \sigma) = \alpha(\lambda \cdot \sigma).$$

It follows that the action of Clifford multiplication globalizes to give an action

$$(Cl(P) \otimes \mathbf{C}) \otimes S_{\mathbf{C}}(\tilde{P}) \to S_{\mathbf{C}}(\tilde{P})$$

which on each fiber is isomorphic to Clifford multiplication. In the case when n is even the action of $Cl_0(P) \otimes \mathbf{C}$ preserves the splitting of $S_{\mathbf{C}}(\tilde{P})$ into $S_{\mathbf{C}}^{\pm}(\tilde{P})$ whereas the action of $Cl_1(P) \otimes \mathbf{C}$ switches the factors.

The Case of the Tangent Bundle and the Cotangent Bundle. Let X be a riemannian manifold and let $P \to X$ be the $SO(n)$-frame bundle associated to the tangent bundle. Then we have the bundle of Clifford algebras $Cl(X)$ associated to P. By definition the bundle of Clifford algebras $Cl(X)$ is constructed from the tangent spaces to X, and thus is a new algebra structure on the exterior algebra of the tangent bundle. Of course, since X is a riemannian manifold there is a canonical identification of the tangent bundle and the cotangent bundle. Thus, we can view $Cl(X)$ as being a new algebra structure on the exterior algebra of the cotangent bundle. We extend this identification complex linearly of the complexified Clifford algebras associated with the tangent and cotangent bundle.

Now suppose that $\tilde{P} \to X$ is a $Spin^c$ structure for X ; i.e., for $P \to X$. We have an action of the bundle $Cl(X)$ on the spin bundle $S_{\mathbf{C}}(\tilde{P})$ by Clifford multiplication. This determines an action of the space of differential forms by Clifford multiplication. The duality of the Clifford multiplication action of one-forms and vector fields is expressed as follows. If α is a complex valued one-form then the action of α on a spinor field s at a point u is given by

$$\alpha \cdot s(u) = \sum_i \alpha(e_i) e_i \cdot s(u). \tag{3.1}$$

where $\{e_1, \ldots, e_n\}$ is an orthonormal basis for TX_u.

Examples. (i) Consider the two-torus T^2 endowed with the standard flat metric. Its tangent bundle and cotangent bundle are trivial. We give these bundles a spin structure. Such a spin structure is a flat S^1-bundle with holonomy in ± 1 and is determined by a representation $\pi_1(T^2) \to \mathbf{Z}/2\mathbf{Z}$. The plus spin bundle is a flat complex line bundle with same holonomy. The minus spin bundle is the inverse complex line bundle.

(ii) Let Σ be a closed oriented riemann surface of genus g. Since $Spin(2) \to SO(2)$ is the double covering of $S^1 \to S^1$, a spin structure for the tangent bundle is equivalent to the choice of a complex line bundle $\mathcal{L}$ of degree $1 - g$ whose square is isomorphic to the tangent bundle. The associated minus spin bundle is this complex line bundle (which is a square root $\sqrt{K_X^{-1}}$ of the canonical line bundle for the riemann surface) and the associated plus spin bundle is its inverse, which is a square root for the canonical line bundle K_X.

3.2   Connections and Curvature

Connections on Principal Bundles.   Let $P \to X$ be a smooth principal G-bundle over a smooth manifold. At each point $p \in P$ we have the vertical tangent space $T^v P_p$. This is the subspace of the tangent space of P which is tangent to the fiber of the projection mapping. A *connection* on this bundle is a distribution $\{H_p\}_{p \in P}$, i.e., a smoothly varying family of linear subspaces of the tangent bundle TP which is everywhere complementary to the vertical distribution and which is invariant under the action of the group G. The condition that the distribution be complementary to the vertical distribution simply means that under the projection mapping each linear subspace of the distribution projects isomorphically onto the tangent space to X at the image point. This condition is expressed by calling the distribution the *horizontal distribution*.

Given a connection on P there is an associated one-form ω on P with values in the adjoint bundle ad P of P; i.e., the vector bundle associated to P and the adjoint action of G on its Lie algebra g. This one-form is called the *connection one-form*. Its value at any $p \in P$ is a linear map ω_p defined as follows:

$$\omega_p : TP_p \to T^v P_p \cong g$$

where the first map is the projection with kernel the subspace H_p, and the second map is the inverse of the isomorphism induced by the action of G at p. The equivariance property of the distribution translates into the condition that ω transforms by the adjoint action: for any $h \in G$, for any $p \in P$, and for any $\tau \in TP_p$ we have

$$\omega_{ph}(\tau \cdot h) = h^{-1}\omega_p(\tau)h.$$

Also, the restriction of ω_p to the fiber of the projection is identified with the left invariant Mauer-Cartan form on G. This is the form on G with values in g whose value at any $\tau \in TG_h$ is equal to $h^{-1}\tau \in g$. It is the unique form on G which is invariant under left multiplication by any element in the group and which is the the identity map at the identity of the group.

These two properties characterize the connection one-forms. Given a connection one-form, one can recover the horizontal distribution as the kernels of the one-form.

Given a connection on a principal bundle P there is an induced connection on any vector bundle $E = P \times_G V$ coming from a linear representation of G on a vector space V. The natural way to view this connection is as a covariant derivative

$$\nabla : \Omega^0(X; E) \to \Omega^1(X; E)$$

which is linear over the scalars ($\mathbf{R}$ or $\mathbf{C}$) and which is a derivation over the scalar-valued functions with respect to the usual d; for any section $\sigma \in \Omega^0(X; E)$ and any scalar-valued function f on X we have

$$\nabla(f \cdot \sigma) = f \cdot \nabla(\sigma) + df \otimes \sigma.$$

The definition of this covariant derivative is given as follows. Suppose that we have a local section σ of E near $x \in X$, and suppose that $\tau \in TX_x$. We choose a smooth curve γ in X through x with tangent τ, and we lift γ to a horizontal curve $p(t)$ in P, i.e., to a curve in P whose tangent vector at each point is contained in the horizontal distribution. Then the restriction of σ to $E|_\gamma$ is expressed as $[p(t), v(t)]$ for some smooth function $v(t)$ with values in V. We define

$$\nabla(\sigma)(\tau) = \left[p, \frac{\partial v}{\partial t} \Big|_{t=0} \right].$$

It is an easy exercise that the result is a well-defined derivation over the functions.

If G is embedded as a subgroup of $Gl(V)$ then one can recover the original connection on the principal G-bundle from the covariant derivative on E. For example, a covariant derivative on a vector bundle associated to a principal $SO(n)$-bundle by the standard representation of $SO(n)$ on $\mathbf{R}^n$ comes from a connection on the principal $SO(n)$ bundle if and only if the covariant derivative ∇ satisfies:

$$d < \sigma_0, \sigma_1 > \, = \, < \nabla(\sigma_0), \sigma_1 > + < \sigma_0, \nabla(\sigma_1) >$$

for all local sections σ_0, σ_1 of the vector bundle. On the other hand, the covariant derivative associated with a trivial representation is the usual derivative on the product bundle and hence reflects no information about the connection on the principal bundle.

Using the derivation rule, we extend ∇ to a map

$$\nabla : \Omega^i(X; E) \to \Omega^{i+1}(X; E)$$

by the formula

$$\nabla(\omega \otimes \sigma) = d\omega \otimes \sigma + (-1)^{\deg \omega} \omega \wedge \nabla(\sigma)$$

for any scalar-valued i-form ω and any section σ of E.

Formulae with respect to a local trivialization for the connection one-form and the covariant derivative. Let $P \to X$ be a principal G-bundle with a connection $\omega \in \Omega^1(P; g)$. We fix a local trivialization of $P|_U$ for some open subset $U \subset X$. We can view this trivialization as a section $\sigma_0: U \to P|_U$. We then have the form $\sigma_0^*(\omega) \in \Omega^1(U; g)$. This is the connection one-form with respect to the given trivialization. If $G \subset Gl(n, \mathbf{R})$ so that $g \subset M_{n \times n}(\mathbf{R})$, then $\sigma_0^*(\omega)$ is a matrix valued one-form $(\widetilde{\omega}_{i,j})$ whose value at every point belongs to the subspace g. For example, if $G = SO(n)$, then $(\widetilde{\omega}_{i,j})$ is a skew-symmetric matrix-valued one-form; i.e., $\widetilde{\omega}_{i,j} = -\widetilde{\omega}_{j,i}$. Suppose that $\sigma: U \to P|_U$ is another section. Then there is a smooth map $h: U \to G$ such that $\sigma(u) = \sigma_0(u)h(u)$. It follows easily from the above descriptions that

$$\sigma^*(\omega)(u) = h(u)^{-1}\sigma_0^*(\omega)(u)h(u) + h(u)^{-1}dh(u).$$

Said another way, if $\sigma(u) = (u, h(u))$ with respect to the original trivialization and if τ is a tangent vector to U at u, then

$$\omega\left(\frac{\partial \sigma}{\partial \tau}\right) = h(u)^{-1}\widetilde{\omega}(\tau)h(u) + h(u)^{-1}\frac{\partial h}{\partial \tau}(u).$$

Now let us suppose that we have a representation $\rho: G \to Gl(V)$ inducing a representation $d\rho: g \to \mathrm{End}(V)$. Let $E \to X$ be the induced vector bundle $P \times_G V$. The trivialization of $P|_U$ induces one for $E|_U$: say $E|_U = U \times V$. We compute ∇ with respect to this trivialization. It becomes a map

$$\nabla: \Omega^0(U; V) \to \Omega^1(U; V).$$

Let $\alpha: U \to V$ be a smooth map determining the section

$$\sigma(u) = (u, \alpha(u))$$

of $E|_U$ in the given trivialization. Then

$$\nabla(\sigma) = d\rho(\widetilde{\omega})(\alpha) + d\alpha.$$

The Curvature of a Connection. If ω is a one-form on P with values in g, then let us denote by $\frac{1}{2}\omega \wedge \omega$ the two-form whose value on a pair of tangent vectors (τ_1, τ_2) and a point p is given by

$$[\omega_p(\tau_1), \omega_p(\tau_2)].$$

Notice that there is an identification between, on the one hand, k-forms on P with values in g which vanish on all k-tuples of tangent vectors at least one of which is vertical and which transform via the adjoint action under the multiplication by G, and on the other hand, k-forms on X with values in $\text{ad}\,P$.

Lemma 3.2.1 *Let ω be a connection one-form on a principal G-bundle P over X. Then the two-form*

$$d\omega + \frac{1}{2}\omega \wedge \omega$$

is a two-form on P which is induced via pullback from a two-form on X with values in $\text{ad}\,P$.

The proof is an easy exercise.

This two-form on X with values in $\text{ad}\,P$ is called the *curvature form* of the connection.

Suppose that $U \subset X$ is an open set and suppose that we have a trivialization of $P|_U$. Any connection one-form ω when restricted to U becomes an untwisted one-form with values in the Lie algebra of G. Let $E \to X$ be a vector bundle associated to P and a representation $\rho: G \to GL(\mathbf{R}^n)$. We have the differential $d\rho$ which is a representation of the Lie algebra of G into $n \times n$ matrices. Furthermore, the trivialization of $P|_U$ induces a trivialization of $E|_U$ with basis $(e_1, \ldots, e_n)$ coming from the standard basis of $\mathbf{R}^n$. The connection on $E|_U$ is then given by

$$\nabla(e_i) = \sum_j (d\rho(\omega)_{j,i}) \otimes e_j.$$

The curvature form is given by

$$\Omega(e_i) = \sum_j \Omega_{j,i} \otimes e_j$$

where

$$\Omega_{j,i} = d\omega_{j,i} + \sum_k \omega_{j,k} \wedge \omega_{k,i}.$$

In the case that $G = SO(n)$, the curvature $(\Omega_{i,j})$ is a skew-symmetric matrix of two-forms; i.e., $\Omega_{i,j} = -\Omega_{j,i}$.

A direct computation shows that for the covariant derivative ∇ in an vector bundle E associated to a principal G-bundle P over X and a representation of G on a vector space V, we have that the operator

$$\nabla \circ \nabla: \Omega^0(X; E) \to \Omega^2(X; E)$$

is linear over the functions and hence is a section of

$$\Omega^2(X; \mathrm{End}(E)).$$

One can show directly that this section is simply the image of the curvature of the connection under the map $g \to \mathrm{End}(V)$ induced by the action of G on V. Indeed, using the local trivializations above we have

$$
\begin{aligned}
\nabla \circ \nabla(e_i) &= \nabla \left(\sum_j \omega_{j,i} \otimes e_j \right) \\
&= \sum_j d\omega_{j,i} \otimes e_j - \sum_j \omega_{j,i} \wedge \nabla(e_j) \\
&= \sum_j d\omega_{j,i} \otimes e_j - \sum_k \omega_{k,i} \wedge \left(\sum_j \omega_{j,k} \otimes e_j \right) \\
&= \sum_j d\omega_{j,i} \otimes e_j + \sum_j \left(\sum_k \omega_{j,k} \wedge \omega_{k,i} \right) \otimes e_j \\
&= \sum_j \left(d\omega_{j,i} + \frac{1}{2}(\omega \wedge \omega)_{j,i} \right) \otimes e_j \\
&= \sum_j \Omega_{j,i} \otimes e_j.
\end{aligned}
$$

What this means more explicitly is that

$$\nabla_{e_r} \circ \nabla_{e_s} - \nabla_{e_s} \circ \nabla_{e_r} = \Omega(e_r, e_s) = \sum_{i,j} \Omega_{j,i}(e_r, e_s) e_j \otimes e_i^*$$

as sections of the endomorphism bundle of $E|_U$. (Here $\Omega_{j,i}$ is a real-valued two-form on X and $\Omega_{j,i}(e_r, e_s)$ denotes the value of this two-form on the ordered pair of tangent vectors (e_r, e_s).)

Out of the curvature form one can construct ordinary closed scalar-valued forms. These are obtained by applying homogeneous polynomials on the Lie algebra which are invariant under the adjoint action of the Lie algebra. A simple exercise shows that the resulting forms are closed and vary by an exact form as one changes the connection. Thus, cohomology classes that they represent are independent of the connection, and all called *characteristic classes* of the bundle. The real Chern classes and the Pontrjagin classes are obtained in this way. For more details, see for example [4] or [16].

There is more than one connection form on a principal bundle $\pi: P \to X$. In fact, for any connection one-form ω and for any one-form η on X with values in $\mathrm{ad}P$, the sum $\omega + \pi^*\eta$ is a connection one-form. In this way one can obtain all connections one-forms on P, so that the space of connection one-forms for P becomes an affine space associated to the vector space $\Omega^1(X; \mathrm{ad}\, P)$.

Examples. (i) Let $P \to X$ be a $U(1)$-bundle. The Lie algebra of $U(1)$ is the subset of purely imaginary complex numbers $i\mathbf{R} \subset \mathbf{C}$. A connection one-form for P is a form $\omega \in \Omega^1(P; i\mathbf{R})$ which is invariant under the $U(1)$-action and which has the property that for any $p \in P$ the pullback of ω to $U(1)$ via the map $U(1) \to P$ given by $h \mapsto ph$ is the form $id\theta$. The two-form $d\omega$ on P is pulled up from a two-form Ω on X with values in $i\mathbf{R}$. This form Ω is the curvature form of the connection. The first Chern class of the associated complex line bundle is by definition represented by the closed form $\frac{i}{2\pi}\Omega$. If we view the group as $SO(2)$ instead of $U(1)$, then its Lie algebra is skew-symmetric two-by-two matrices. The generator i of the Lie algebra of $U(1)$ corresponds to the element $e_1 \wedge e_2$ which is the matrix

$$\begin{pmatrix} 0 & -1 \\ 1 & 0 \end{pmatrix}.$$

In this notation the connection one-form is given by

$$\begin{pmatrix} 0 & -\omega/i \\ \omega/i & 0 \end{pmatrix},$$

and its curvature is

$$\begin{pmatrix} 0 & -\Omega/i \\ \Omega/i & 0 \end{pmatrix}.$$

By definition, the Gaussian curvature is the $(1,2)$-entry of the curvature matrix. Thus, the Gaussian curvature is $-\Omega/i = i\Omega$. This agrees with the usual Gauss-Bonnet theorem which tells us that $\frac{1}{2\pi}$ times the Gaussian curvature integrates to give the Euler characteristic which is also the integral of c_1 of the $U(1)$-bundle.

Action of the Group of Changes of Gauge. A *change of gauge* is simply a bundle automorphism of the principal bundle. Let us consider the action of the group of automorphisms of $P \to X$ on the space of connection one-forms. If $\varphi: P \to P$ is a diffeomorphism commuting with the action of G and with the projection to X, then φ is said to be an automorphism of P. These form a group, denoted $\mathrm{Aut}(P)$ under composition acting on the left on P commuting with the right action of G on P. We can interpret such a bundle automorphism as a function from P to G which satisfies $\psi(ph) = h^{-1}\psi(p)h$ for any $h \in G$. The relationship is that the function $\psi\colon P \to G$ corresponds to the bundle automorphism $\varphi(p) = p\psi(p)$. Clearly, any function defined this way commutes with the projection to X. The equation for ψ above is equivalent to the statement that φ is equivariant with respect to the G-action.

If ω is a connection one-form and φ is a bundle automorphism then, $\varphi^*\omega$ is also a connection one-form. It turns out that

$$\varphi^*(\omega) = \varphi^{-1}\omega\varphi + \varphi^{-1}d\varphi.$$

Maybe the easiest way to understand this equation is to consider it as an equation of g-valued one-forms on P.

The effect of this action on the curvature of the connection is to conjugate it by φ; i.e., if Ω is the curvature form for the connection ω and Ω' is that for $\varphi^*\omega$, then

$$\Omega' = \varphi^{-1}\Omega\varphi.$$

In particular, the norm-square of the curvature is left invariant by the action of the group of changes of gauge.

The Levi-Città Connection.   Now suppose that $P \to X$ is the principal $SO(n)$-bundle underlying the tangent bundle of a riemannian manifold. As we have seen above there are many connections on the tangent bundle which are compatible with this principal bundle; i.e., come from $SO(n)$ connections on P. It turns out that there is a best one, the Levi-Città connection.

To introduce this connection, let us consider a local coordinate system $U \subset X$ with local coordinates $\{x^1, \ldots, x^n\}$. Let e_i be the tangent vector $\partial/\partial x^i$, and let the metric be given by the symmetric matrix $\{g_{i,j}\}$ with respect to this coordinate system. Let ∂_i denote ordinary differentiation with respect to x^i. Then for any covariant derivative there are n^3 functions $\Gamma^k_{i,j}$, the Christoffel symbols, such that

$$\nabla_{e_i}(e_j) = \sum_{k=1}^{n} \Gamma^k_{i,j} e_k.$$

Then the condition that this connection is an orthogonal connection is that

$$\langle \nabla_{e_k}(e_i), e_j \rangle + \langle e_i, \nabla_{e_k}(e_j) \rangle = \partial_k(\langle e_i, e_j \rangle).$$

In terms of Christoffel symbols this means

$$\partial_k g = \Gamma_k \cdot g + g \cdot (\Gamma_k)^{tr}$$

where Γ_k is the matrix whose (r, s)-entry is $\Gamma^s_{k,r}$. The extra condition that picks out the Levi-Cività connection is that the connection be torsion-free. This is a condition that can only be defined in the case of connections on the tangent bundle. It says that for any pair of vector fields V, W on X we have

$$\nabla_V(W) - \nabla_W(V) = [V, W].$$

In the case when V and W are partial derivatives in a local coordinate system, the bracket term vanishes and the torsion-free condition becomes

$$\nabla_{e_i}(e_j) = \nabla_{e_j}(e_i).$$

In terms of the Christoffel symbols, this means that

$$\Gamma^j_{k,i} = \Gamma^j_{i,k}$$

for all i, j, k. That is to say the (i, j) entry of Γ_k is equal to the (k, j) entry of Γ_i.

Lemma 3.2.2 *There is a unique torsion-free orthogonal connection on the tangent bundle of a riemannian manifold.*

Proof. Let us first prove uniqueness. Suppose that $\Gamma^j_{k,i}$ and $\hat{\Gamma}^j_{k,i}$ are two sets of Christoffel symbols giving torsion-free orthogonal connections in a local coordinate system. For each k form the difference matrix $\mu_k = \Gamma_k - \hat{\Gamma}_k$. The fact that both the Christoffel symbols are orthogonal connections yields that for all k we have

$$\mu_k \cdot g + (\mu_k \cdot g)^{tr} = 0.$$

The fact that both connections are torsion-free gives.

$$(\mu_k)_{i,j} = (\mu_i)_{k,j}.$$

Now let us set $\nu_k = \mu_k \cdot g$. The first equation tells us that for all k we have

$$\nu_k + \nu_k^{tr} = 0.$$

The second equation yields

$$(\nu_k)_{i,j} = (\nu_i)_{k,j}$$

for all i, j, k. Thus, we have

$$\nu^{j}_{k,i} = -\nu^{i}_{k,j} = -\nu^{i}_{j,k} = \nu^{k}_{j,i} = \nu^{k}_{i,j} = -\nu^{j}_{i,k} = -\nu^{j}_{k,i}.$$

Thus, we see that $\nu^{j}_{k,i} = 0$ for all i, j, k. Since g is invertible, this proves that $\mu_k = 0$ for all k, which implies that $\Gamma_k = \hat{\Gamma}_k$ for all k. This proves uniqueness.

As to existence, for each k we define

$$(\Upsilon_k)_{i,j} = \frac{1}{2}\left(\partial_k g_{i,j} + \partial_i g_{j,k} - \partial_j g_{k,i}\right)$$

We then define

$$\Gamma_k = \Upsilon_k \cdot g^{-1}.$$

A direct computation shows that these Christoffel symbols define an orthogonal, torsion-free connection. Another way to prove the existence of the Levi-Cività connection is to choose at $x \in X$ a local coordinate system so that the metric is standard to second order at the point (local Gaussian coordinates). In these coordinates the Christoffel symbols vanish at the origin.   □

Definition 3.2.3 The Levi-Cività connection is called the *riemannian connection* for the riemannian manifold.

Connections on Spin Bundles. Let $P \to X$ be an arbitrary $SO(n)$-bundle and suppose that $\tilde{P} \to X$ is a lifting of P to a spin bundle. Then any $SO(n)$-connection on P automatically lifts to a $Spin(n)$ connection on $\tilde{P}$. Thinking of the connection as a horizontal distribution in P, this is completely clear. The space $\tilde{P}$ is a two-sheeted covering of P, and we simply lift the horizontal distribution via the inverse of the differential of the covering map at each point. This is the unique $Spin(n)$ connection which projects to the given $SO(n)$ connection.

In the case that $P \to X$ is the orthogonal frame bundle of the tangent bundle and the connection is the riemannian connection, we call the induced connection on $\tilde{P}$ the *spin connection*.

Let us return to the case of an arbitrary $SO(n)$-bundle $P \to X$ endowed with a connection and a spin lifting $\tilde{P}$ for P. The connection on P induces one on $Cl(P)$. The lifted connection on $\tilde{P}$ induces one on $S_{\mathbf{C}}(\tilde{P})$. We denote the covariant derivatives of these connections by ∇ and $\tilde{\nabla}$ respectively. The connection $\tilde{\nabla}$ is a unitary connection on $S_{\mathbf{C}}(\tilde{P})$ since $Spin(n)$

acts by unitary transformations under the spin representation $\Delta_{\mathbf{C}}$. The connection of $Cl(P)$ is a derivation with respect to the algebra structure in the sense that

$$\nabla(\lambda_1 \lambda_2) = \nabla(\lambda_1)\lambda_2 + \lambda_1 \nabla(\lambda_2)$$

for any sections λ_1, λ_2 of $Cl(P)$. Also, the action of $Cl(P)$ on $S_{\mathbf{C}}(\widetilde{P})$ is then compatible with these connections in the sense that

$$\widetilde{\nabla}(\lambda \cdot \sigma) = \nabla(\lambda) \cdot \sigma + \lambda \cdot \widetilde{\nabla}(\sigma)$$

for any sections λ of $Cl(P)$ and σ of $S_{\mathbf{C}}(\widetilde{P})$.

It is easy to see that the splitting $Cl(P) = Cl_0(P) \oplus Cl_1(P)$ is invariant under ∇, and hence these two sub-bundle receive induced connections and covariant derivatives. Also, the complex unit $\omega_{\mathbf{C}}$ is self-parallel under the connections in the sense that $\nabla(\omega_{\mathbf{C}}) = 0$. This implies that $(Cl(P) \otimes \mathbf{C})^{\pm}$ are invariant under the covariant derivative. In the case of $SO(2n)$-bundles P the splitting of $S_{\mathbf{C}}(\widetilde{P})$ into $S_{\mathbf{C}}^{\pm}(\widetilde{P})$ is also invariant under the connection, so that both of these bundles receive unitary connections which we denote $\widetilde{\nabla}$.

The Lie Algebra of $Spin(n)$ and its action on $Cl(\mathbf{R}^n)$. The double covering map $Spin(n) \to SO(n)$ identifies the Lie algebra of $Spin(n)$ with the Lie algebra $so(n)$ of $SO(n)$. That is to say, the Lie algebra of $Spin(n)$ is the space of skew-symmetric $n \times n$ real matrices. A basis for this Lie algebra is $e_i \wedge e_j$ for $1 \leq i < j \leq n$, where $e_i \wedge e_j$ denotes the matrix tangent at the identity to the one parameter subgroup defined by sending θ to a rotation through angle θ in the i, j plane, moving from e_i towards e_j. Explicitly, this matrix has a -1 in the (i, j) place, a 1 in the (j, i) place, and zero in all other places.

Lemma 3.2.4 *Using the natural identification of the Lie algebras of $SO(n)$ and $Spin(n)$, the infinitessimal generator for the one parameter subgroup*

$$\theta \mapsto \cos(\theta)1 + \sin(\theta)e_i e_j \subset Spin(n)$$

is $2(e_i \wedge e_j)$.

Proof. First notice that the given elements are expressed as products $e_i(-\cos(\theta)e_i + \sin(\theta)e_j)$ so that this is a subgroup of $Spin(n)$. The natural map from $Spin(n) \to SO(n)$ induced by the conjugation action of $Spin(n)$ on $\mathbf{R}^n \subset Cl(\mathbf{R}^n)$ induces the identification of the Lie algebras. Thus, to compute the infinitessimal generator for the given one-parameter subgroup, we compute the infinitessimal generator for the induced one-parameter sub-group of rotations of $\mathbf{R}^n$. Since the conjugation action of a unit vector v on $\mathbf{R}^n$ is minus the reflection R_v in the plane perpendicular to v we see that the conjugation action of the given element on $\mathbf{R}^n$ is

$$R_{e_i} \circ R_v$$

where

$$v = -\cos(\theta)e_i + \sin(\theta)e_j.$$

This composition is simply rotation in the (i, j) plane through an angle 2θ from e_i toward e_j. Hence, its infinitessimal generator is $2e_i \wedge e_j$. $\square$

Of course, viewing the one parameter subgroup given in the previous lemma as a submanifold in $Cl(\mathbf{R}^n)$ the differential at the identity is $e_1 e_2$. Thus, we have established that under the differential of the double covering map from $Spin(n) \to SO(n)$ at the identity the element $e_i e_j \in Cl(\mathbf{R}^n)$ maps to $2(e_i \wedge e_j) \in so(n)$.

Explicit formula for the connections on $S_\mathbf{C}(\widetilde{P})$.   Let $P \to X$ be an $SO(n)$ bundle with connection ω, and let $\widetilde{P} \to X$ be a lifting of P to a spin bundle. Let

$$\widetilde{\nabla}: \Omega^0(X; S_\mathbf{C}(\widetilde{P})) \to \Omega^1(X; S_\mathbf{C}(\widetilde{P}))$$

be the induced covariant derivative. It is our purpose here to fix a local trivialization of $P \to X$ (and hence of $S_\mathbf{C}(\widetilde{P})$) and to derive a local formula for $\widetilde{\nabla}$ in terms of the connection matrix $\widetilde{\omega} = (\widetilde{\omega}_{i,j})$ for ω.

Fix an open subset $U \subset X$ and a trivialization $P|_U = U \times SO(n)$. Suppose that with respect to these choices the connection $\omega|_U$ is given by a matrix-valued one-form $\widetilde{\omega}_{i,j}$ on U. Of course, since ω is an orthogonal connection, the matrix $(\widetilde{\omega}_{i,j})$ is skew-symmetric. Since we have identified the Lie algebras of $SO(n)$ and $Spin(n)$ via the double covering map, for the induced trivialization for $\widetilde{P}|_U$ the induced connection is given by the same matrix of one-forms $(\widetilde{\omega}_{i,j})$.

Let $\sigma(u)$ be a section of $S_\mathbf{C}(\widetilde{P}|_U)$. Using the given trivialization we write $\sigma(u) = (u, s(u))$. Then we have

$$\widetilde{\nabla}(\sigma)(u) = (u, d\rho(\widetilde{\omega}_{i,j})(s(u)) + ds(u))$$

where $d\rho$ is the action of $so(n)$ on $S_\mathbf{C}(\mathbf{R}^n)$ induced from the action of $Spin(n)$ on $S_\mathbf{C}(\mathbf{R}^n)$. Of course, since $e_i \wedge e_j$ is the skew-symmetric matrix with -1 in the (i, j) place and $+1$ is the (j, i) place we have that the matrix $(\widetilde{\omega}_{i,j})$ is the element

$$\sum_{i<j} \widetilde{\omega}_{j,i} e_i \wedge e_j$$

in $so(n)$. According to Lemma 3.2.4 the element $e_i \wedge e_j$ corresponds to $(e_i e_j)/2 \in Cl(\mathbf{R}^n)$. Thus,

$$d\rho(\widetilde{\omega}_{i,j}) = \frac{1}{2} \sum_{i<j} \widetilde{\omega}_{j,i} e_i e_j$$

where on the right-hand-side the action on $S_{\mathbf{C}}(\mathbf{R}^n)$ is Clifford multiplication. Thus, the final formula written out with respect to the given trivialization is

$$\widetilde{\nabla}(\sigma)(u) = \left(u, ds(u) + \frac{1}{2} \sum_{i<j} \widetilde{\omega}_{j,i} e_i e_j \cdot s(u) \right)$$

where σ is the section $\sigma(u) = (u, s(u))$.

Connections on $Spin^c(n)$-bundles. Now let $P \to X$ be an $SO(n)$-bundle with a connection and let $\widetilde{P} \to X$ be a lifting to a $Spin^c$-connection. In this case, since $Spin^c(n) \to SO(n)$ is not a finite covering, we cannot automatically lift a connection ω on P to one on $\widetilde{P}$. We need another piece of information; namely a $U(1)$-connection A on the determinant line bundle $\mathcal{L} \to X$ of $\widetilde{P}$. Given ω and A, this determines a connection on the principal $SO(n) \times S^1$-bundle which is the quotient of $\widetilde{P}$ by $\{\pm 1\} \subset Spin^c(n)$. As before, there is a unique lifting of this connection to one on $\widetilde{P}$. Suppose that we fix a local trivialization for P and for the determinant line bundle $\mathcal{L}$, and that with respect to these local trivializations the connections are given by $(\widetilde{\omega}_{i,j})$ and iA respectively. Of course, $S_{\mathbf{C}}(\widetilde{P})$ receives an induced local trivialization. In this local trivialization covariant differentiation is given by

$$\widetilde{\nabla}_e(\sigma)(u) = \left(u, \frac{ds(u)}{de} + \frac{1}{2} \left(iA(e) + \sum_{i<j} \widetilde{\omega}_{j,i}(e) e_i e_j \right) \cdot s(u) \right) \qquad (3.2)$$

where σ is the section given by $u \mapsto (u, s(u))$ in the induced local trivialization of $\widetilde{P}$.

3.3   The Dirac Operator

In this paragraph we shall introduce the Dirac operator on the spin bundles and $Spin^c$ bundles of a riemannian manifold. We shall also establish some of the basic properties of this operator.

Let X be a riemannian manifold; let $P \to X$ be the $SO(n)$-principal bundle associated to the tangent bundle; and let $\widetilde{P}$ be a lifting of this bundle to a spin or $Spin^c$-bundle. Let $S_{\mathbf{C}}(\widetilde{P})$ be the associated spin bundle. In the case of a $Spin^c$ lifting let us also fix a $U(1)$-connection A on the determinant line bundle. Let $\widetilde{\nabla}$ be the spin connection induced by the Levi-Cività connection and the connection A. We define the Dirac operator

$$\displaystyle{\not{\partial}}_A : C^\infty(S_{\mathbf{C}}(\widetilde{P})) \to C^\infty(S_{\mathbf{C}}(\widetilde{P}))$$

as follows:

$$\displaystyle{\not{\partial}}_A(\sigma)(x) = \sum_{i=1}^n e_i \cdot \widetilde{\nabla}_{e_i}(\sigma)(x)$$

where $\{e_1, \ldots, e_n\}$ is an oriented orthonormal frame for TX_x and where the $\cdot$ is Clifford multiplication. In the case of a $Spin$ bundle there is no connection A and we denote the Dirac operator by $\not{\partial}$.

Lemma 3.3.1 *The operators ${\not{\partial}}_A$ and $\not{\partial}$ defined above are independent of the choice of orthonormal frame $\{e_1, \ldots, e_n\}$.*

Proof.   Suppose that $\{e'_1, \ldots, e'_n\}$ is another oriented orthonormal frame. Suppose that

$$e'_i = \sum_{j=1}^n B_{i,j} e_j.$$

Then B is an element of $SO(n)$. Let us consider

$$\sum_{i=1}^n e'_i \cdot \widetilde{\nabla}_{e'_i}(\sigma).$$

Since Clifford multiplication is bilinear we see that

$$e'_i \cdot \alpha = \sum_{j=1}^n B_{i,j} e_j \cdot \alpha$$

for any $\alpha \in S_{\mathbf{C}}(\widetilde{P})$. Also, since $\widetilde{\nabla}_e$ is linear in e we see that

$$\widetilde{\nabla}_{e'_i}(\sigma) = \sum_{j=1}^n B_{i,j} \widetilde{\nabla}_{e_j}(\sigma).$$

Combining these we have

$$\sum_{i=1}^{n} e_i' \cdot \tilde{\nabla}_{e_i'}(\sigma) = \sum_{i=1}^{n} \left(\sum_{j=1}^{n} B_{i,j} e_j \cdot \left(\sum_{j'=1}^{n} B_{i,j'} \tilde{\nabla}_{e_{j'}}(\sigma) \right) \right)$$

$$= \sum_{i,j,j'} B_{i,j} B_{i,j'} e_j \cdot \tilde{\nabla}_{e_{j'}}(\sigma).$$

Since B is an orthogonal matrix we have

$$\sum_i B_{i,j} B_{i,j'} = \delta_{j,j'}.$$

Thus,

$$\sum_{i=1}^{n} e_i' \cdot \tilde{\nabla}_{e_i'}(\sigma) = \sum_{j,j'} \delta_{j,j'} e_j \cdot \tilde{\nabla}_{e_{j'}}(\sigma)$$

$$= \sum_{j=1}^{n} e_j \cdot \tilde{\nabla}_{e_j}(\sigma).$$

This completes the proof. $\qquad\qquad\qquad\square$

Let us write out the expression for the Dirac operator with respect to a local trivialization of the principal $SO(n)$-bundle $P \to X$ associated to the tangent bundle of a riemannian manifold X with a spin structure $\tilde{P}$. Let $(\tilde{\omega}_{i,j})$ be the connection one-form in the local trivialization. Let σ be a local section of $S_{\mathbf{C}}(\tilde{P})$ given by $\sigma(u) = (u, s(u))$ in the induced local trivialization. Let $e_1, \ldots, e_n$ be the orthonormal basis at $u \in X$ corresponding to the standard basis for $\mathbf{R}^n$ under the trivialization. We have

$$\partial\!\!\!/(\sigma)(u) = \sum_i e_i \tilde{\nabla}_{e_i}(\sigma)(u)$$

$$= \left(u, \sum_i e_i \left(\frac{ds(u)}{de_i} + \frac{1}{2} \sum_{j<k} \tilde{\omega}_{k,j}(e_i)(e_j e_k) \cdot s(u) \right) \right)$$

$$= \left(u, \sum_i e_i \cdot \frac{ds(u)}{de_i} + \frac{1}{2} \sum_i \sum_{j<k} \tilde{\omega}_{k,j}(e_i)(e_i e_j e_k) \cdot s(u) \right),$$

where the $\cdot$ represents Clifford multiplication. One sees immediately from the last expression that the operator $\partial\!\!\!/$ at the point p is of the form $\sum_i e_i \cdot \frac{\partial}{\partial e_i}$

plus a zeroth order term (one involving no derivatives). In particular, $\partial\!\!\!/$ is a linear, first-order operator.

In the case of a *Spin^c* lifting $\widetilde{P}$ and a $U(1)$-connection A on $\det(\widetilde{P})$, the formula is

$$\partial\!\!\!/_A(\sigma)(u) = \sum_\ell e_\ell \cdot \frac{ds(u)}{de_\ell} + \frac{1}{2}\sum_\ell \left(A(e_\ell)e_\ell + \sum_{j<k} \widetilde{\omega}_{k,j}(e_\ell)(e_j e_k) \right) \cdot s(u).$$
(3.3)

Notice that by Equation 3.1 the expression $\sum_\ell A(e_\ell)e_\ell \cdot s(u)$ is equal to the Clifford product $A \cdot s(u)$.

We also have a simple formula for the how the Dirac operator changes with the connection A.

Lemma 3.3.2 *Let A and $A' = A + \alpha$ be two $U(1)$-connections on the determinant line bundle $\mathcal{L}$ of a Spin^c structure $\widetilde{P}$ for X. Then for any section ψ of $S_{\mathbf{C}}(\widetilde{P})$ we have*

$$\partial\!\!\!/_{A'}(\psi) = \partial\!\!\!/_A(\psi) + \frac{1}{2}\alpha \cdot \psi.$$

Proof.   This is clear from Equation 3.3 and the remark following the equation.   □

Notice that in the case of even dimensional manifolds $\partial\!\!\!/$ and $\partial\!\!\!/_A$ both map $C^\infty(S_{\mathbf{C}}^\pm(\widetilde{P}))$ to $C^\infty(S_{\mathbf{C}}^\mp(\widetilde{P}))$.

There is one special property that $\partial\!\!\!/$ satisfies: it is formally self-adjoint.

Lemma 3.3.3 *Let X be a closed manifold with a spin or Spin^c structure $\widetilde{P}$. Then $\partial\!\!\!/: S_{\mathbf{C}}(\widetilde{P}) \to S_{\mathbf{C}}(\widetilde{P})$ is formally self-adjoint in the sense that*

$$(\partial\!\!\!/(\sigma_1), \sigma_2)_{L^2} = (\sigma_1, \partial\!\!\!/(\sigma_2))_{L^2}$$

where $(\cdot, \cdot)_{L^2}$ denotes the L^2-inner product on sections of $S_{\mathbf{C}}(\widetilde{P})$ induced from the pointwise hermitian inner product on the fibers.

Proof.   The hermitian inner product on C^∞ sections of $S_{\mathbf{C}}(\widetilde{P})$ is given by

$$(\partial\!\!\!/(\sigma_1), \sigma_2)_{L^2} = \int_X \langle \partial\!\!\!/(\sigma_1), \sigma_2 \rangle dvol$$

where the inner product on the right-hand-side is the pointwise hermitian inner product on the fibers of the complex spin bundle. Let us fix a coordinate system at a point x so that for the standard unit tangent vectors $e_1, \ldots, e_n$ we have $\nabla_{e_i}(e_i) = 0$ at the point x. Computing at x we have

$$
\begin{aligned}
\langle \partial\!\!\!/(\sigma_1), \sigma_2 \rangle_x &= \langle \sum_i e_i \tilde{\nabla}_{e_i}(\sigma_1), \sigma_2 \rangle_x \\
&= -\sum_i \langle \tilde{\nabla}_{e_i}(\sigma_1), e_i \sigma_2 \rangle_x \\
&= \sum_i \langle \sigma_1, \tilde{\nabla}_{e_i}(e_i \sigma_2) \rangle_x - \frac{\partial}{\partial e_i} \langle \sigma_1, e_i \sigma_2 \rangle_x \\
&= \sum_i \langle \sigma_1, e_i \tilde{\nabla}_{e_i}(\sigma_2) \rangle_x + \langle \sigma_1, \nabla_{e_i}(e_i)\sigma_2 \rangle_x - \frac{\partial}{\partial e_i} \langle \sigma_1, e_i \sigma_2 \rangle_x \\
&= \sum_i \langle \sigma_1, e_i \nabla_{e_i}(\sigma_2) \rangle_x - \frac{\partial}{\partial e_i} \langle \sigma_1, e_i \sigma_2 \rangle_x \\
&= \langle \sigma_1, \partial\!\!\!/(\sigma_2) \rangle_x - \sum_i \frac{\partial}{\partial e_i} \langle \sigma_1, e_i \sigma_2 \rangle_x.
\end{aligned}
$$

Now let us define a complexified vector field V on X, i.e., a section of $TX \otimes \mathbf{C}$ by the condition

$$
\langle V(x), W(x) \rangle = \langle \sigma_1(x), W \cdot \sigma_2(x) \rangle
$$

for all vector fields W and all $x \in X$. Then the above equalities can be written

$$
\langle \partial\!\!\!/(\sigma_1), \sigma_2 \rangle_x = \langle \sigma_1, \partial\!\!\!/(\sigma_2) \rangle_x - \operatorname{div}(V)_x.
$$

Since all the quantities in this expression are global, in fact the expression holds at every point $x \in X$. Thus, by integrating over X we have

$$
\langle \partial\!\!\!/(\sigma_1), \sigma_2 \rangle_{L^2} = \langle \sigma_1, \partial\!\!\!/(\sigma_2) \rangle_{L^2} - \int_X \operatorname{div}(V) dvol.
$$

The result follows immediately. $\qquad \square$

The Symbol of $\partial\!\!\!/$ and $\partial\!\!\!/_A$. Let D be a first order differential operator from sections of a bundle $E \to X$ to sections of $F \to X$. Then the symbol of D, $\mathrm{Symb}(D)$ is a bundle map from $\pi^* E$ to $\pi^* F$ between the pull backs of the bundles over the cotangent bundle $T^* X$ of X

The symbol of a differential operator is a bundle map which is a function only of its leading order term. In our case, the symbol is a function only of the first order term. Let us fix local coordinates $(x^1, \ldots, x^n)$ which are orthonormal at x. Suppose that an operator is given in local coordinates by

$$D(\sigma) = \sum_I \alpha_I \frac{\partial^{|I|}}{\partial x^I} + \text{lower order terms}$$

where the leading sum ranges over multi-indices I of total length $|I| = n$ and α_I is a linear map from E to F. Then the symbol of D on a cotangent vector $\xi \in T^* X_x$ is the linear map

$$\text{Symb}(D)(\xi) = i^n \sum_I \alpha_I \xi^I$$

where $\xi = (\xi_1, \ldots, \xi_n)$ in the dual basis of $T^* X$ to $(\partial/\partial x^1, \ldots, \partial/\partial x^n)$. Thus, the symbol of $\partial\!\!\!/$ on $T^* X_x$ is given by

$$\text{Symb}(\partial\!\!\!/)(\xi) = i \sum_j \xi_j \cdot (\ )$$

where $\cdot$ is Clifford multiplication. Using the identification of $T^* X_x$ and TX we can view $T^* X_x$ as acting on $S_{\mathbf{C}}(\widetilde{P})$ by Clifford multiplication. The symbol of the Dirac operator is then simply Clifford multiplication by $i\xi$.

For any nonzero cotangent vector ξ Clifford multiplication by $i\xi$ induces an isomorphism from the fiber of $S_{\mathbf{C}}(\widetilde{P})$ to itself. By definition, this means that $\partial\!\!\!/$ is an elliptic first order linear differential operator.

In the same way the symbol of $\partial\!\!\!/_A$ at a cotangent vector ξ is Clifford multiplication by $i\xi$.

Because $\partial\!\!\!/$ and $\partial\!\!\!/_A$ are elliptic, we can say quite a bit about them when the base manifold is compact. We can view them as operators on L_k^2-sections of $S_{\mathbf{C}}(\widetilde{P})$ to L_{k-1}^2-sections of the same bundle. These operators are Fredholm. Their kernels are finite dimensional and are independent of k. In fact, any L^2-section in the kernel of $\partial\!\!\!/$ or $\partial\!\!\!/_A$ is in fact a C^∞-section. Their images are closed subspaces of finite codimension in the space of L_{k-1}^2-sections. Once again, any element orthogonal under the L^2-metric with the image of $\partial\!\!\!/$ or $\partial\!\!\!/_A$ is in fact a C^∞-section.

Examples. Let us begin with a compact connected, oriented riemannian one-manifold X. The metric is automatically flat and hence isometric to a circle of some radius $R > 0$. The orientation determines a trivialization of the orthogonal tangent bundle P_X. Let $\widetilde{P} \to X$ be a spin structure for X. The trivialization of P_X determines a local trivialization of $\widetilde{P}$ whose holonomy is either ± 1, depending on which spin structure we have chosen.

The complex spin bundle $S_{\mathbf{C}}(\widetilde{P})$ is isomorphic to $\widetilde{P} \times_{\pm 1} \mathbf{C}$. This bundle inherits a local trivialization with holonomy either ± 1. Clifford multiplication by a unit tangent vector in the positive direction is multiplication by i. The covariant derivative on $S_{\mathbf{C}}(\widetilde{P})$ is the usual derivation with respect to a local trivialization. Thus, we see that the dirac operator

$$\partial\!\!\!/ : S_{\mathbf{C}}(\widetilde{P}) \to S_{\mathbf{C}}(\widetilde{P})$$

is given by $i\frac{d}{d\theta}$ operating on either the space of complex-valued functions on S^1 (in the case that the holonomy of the flat structure on $\widetilde{P}$ is trivial) or on the space of complex-valued functions f on $\mathbf{R}^1$ which satisfy $f(x + 2\pi R) = -f(x)$ (when the holonomy of the flat structure of $\widetilde{P}$ is -1). In the first case, the kernel and cokernel of $\partial\!\!\!/$ are one-dimensional and consist of the constants sections. In the second case $\partial\!\!\!/$ is an isomorphism.

Now let us consider a closed, oriented riemannian two-manifold X. Let $P_X \to X$ be the orthogonal tangent bundle. It is an S^1-bundle or equivalently a $U(1)$-bundle. The Clifford algebra of $\mathbf{R}^2$ is identified with $\mathbf{H}$ in such a way that $e_1 e_2$ corresponds to i, e_1 to j, and e_2 to k. The group $Spin(2)$ is the unit circle in $\mathbf{C} \subset \mathbf{H}$, i.e., all elements of the form $\cos(\theta) + \sin(\theta)e_1 e_2$. A spin structure $\widetilde{P}$ on P_X is simply a $U(1)$-bundle which is a square root of P_X. We write $\mathbf{H} = \mathbf{C} \oplus j\mathbf{C}$ and the action of $Cl(\mathbf{R}^2) = \mathbf{H}$ is given by right Clifford multiplication. Thus, the action of $Spin(2)$ on $\mathbf{C} \oplus j\mathbf{C}$ is given by

$$\zeta \mapsto \begin{pmatrix} \zeta & 0 \\ 0 & -\zeta \end{pmatrix}$$

for all $\zeta \in S^1$. The elements $e_1 e_2, e_1, e_2$ are given by multiplication by i, j, k respectively. Hence, in the decomposition $\mathbf{C} \oplus j\mathbf{C}$ they are given by the matrices

$$\begin{pmatrix} i & 0 \\ 0 & -i \end{pmatrix}, \begin{pmatrix} 0 & -1 \\ 1 & 0 \end{pmatrix}, \begin{pmatrix} 0 & -i \\ -i & 0 \end{pmatrix}.$$

Thus, $S_{\mathbf{C}}(\widetilde{P})$ is a direct sum of two complex line bundles, one of which is associated to $\widetilde{P}$ by the standard representation of S^1 into $U(1)$ and the other is the inverse $U(1)$-bundle. Since $\omega_{\mathbf{C}} = ie_1 e_2$, $S_{\mathbf{C}}^+(\widetilde{P})$ is the complex line bundle corresponding to the $-i$ eigenspace of $e_1 e_2$, that is to say the inverse of the standard representation. This bundle is $\sqrt{K_X}$, the square root of the canonical bundle K_X of the almost complex structure. The inverse complex line bundle, which is associated to the square root of P_X, is $S_{\mathbf{C}}^-(\widetilde{P})$. This bundle is $\sqrt{K_X^{-1}} = K_X^{-1} \otimes \sqrt{K_X}$. The Levi-Cività connection on P_X induces one on $\widetilde{P}$ and hence covariant derivatives on

$S_C^\pm(\tilde{P})$. Picking an orthonormal basis $\{e_1, e_2\}$ at a point and decomposing $S_C(\tilde{P}) = S_C^+(\tilde{P}) \oplus S_C^-(\tilde{P})$, the dirac operator is given by

$$\begin{pmatrix} 0 & -\nabla_{e_1} - i\nabla_{e_2} \\ \nabla_{e_1} - i\nabla_{e_2} & 0 \end{pmatrix}$$

In particular if the metric is standard to second order at a point in the given coordinate system, then at that point we have

$$\partial\!\!\!/ = \begin{pmatrix} 0 & \frac{\partial}{-\partial e_1} - i\frac{\partial}{\partial e_2} \\ -\frac{\partial}{\partial e_1} - i\frac{\partial}{\partial e_2} & 0 \end{pmatrix}.$$

Note that the decomposition that we are taking when we write these two-by-two matrices is $S_C^+(\tilde{P}) \oplus S_C^-(\tilde{P})$.

The metric and the orientation determine a complex structure on X. There is a local coordinate for this complex structure z such that with respect to the underlying real coordinates (e_1, e_2) the riemannian metric is standard to second order. With respect to a local holomorphic coordinate we have $2\frac{\partial}{\partial \bar{z}} = \frac{\partial}{\partial e_1} + i\frac{\partial}{\partial e_2}$ so that we can rewrite the Dirac operator as

$$\partial\!\!\!/ = \begin{pmatrix} 0 & 2\frac{\partial}{\partial z} \\ -2\frac{\partial}{\partial \bar{z}} & 0 \end{pmatrix}$$

where $-\frac{\partial}{\partial \bar{z}}$ is an operator from the sections of $\sqrt{K_X}$ to the sections of $\sqrt{K_X^{-1}} = \sqrt{K_X} \otimes K_X^{-1}$ and $\frac{\partial}{\partial z}$ is its adjoint.

Now let us consider the case of $\mathbf{R}^3$. The Clifford algebra is isomorphic to $\mathbf{H} \oplus \mathbf{H}$ where the two factors are the plus and minus eigenspace for the action of $\omega_C = -e_1 e_2 e_3$. We take as the spin representation the action of $Cl(\mathbf{R}^3)$ that factors through the plus factor. Thus,

$$S_C(\mathbf{R}^3) = \mathbf{C} \oplus j\mathbf{C} = \mathbf{H}.$$

The basis for the plus factor is

$$\left\{ \frac{1 + \omega_C}{2}, \frac{e_3 + e_1 e_2}{2}, \frac{e_1 + e_2 e_3}{2}, \frac{e_2 + e_3 e_1}{2} \right\}$$

corresponding to $1, i, j, k \in \mathbf{H}$. The matrices corresponding to these elements are

$$\begin{pmatrix} 1 & 0 \\ 0 & 1 \end{pmatrix}, \begin{pmatrix} i & 0 \\ 0 & -i \end{pmatrix}, \begin{pmatrix} 0 & -1 \\ 1 & 0 \end{pmatrix}, \begin{pmatrix} 0 & -i \\ -i & 0 \end{pmatrix}.$$

Letting $\mathbf{R}^2$ be spanned by e_1, e_2, the isomorphism $Cl(\mathbf{R}^2) \to Cl^+(\mathbf{R}^3)$ commutes with these isomorphisms to $\mathbf{H}$. This means that the restriction of $S_{\mathbf{C}}(\mathbf{R}^3)$ to $\mathbf{R}^2$ gives $S_{\mathbf{C}}(\mathbf{R}^2) = S_{\mathbf{C}}^-(\mathbf{R}^2) \oplus S_{\mathbf{C}}^+(\mathbf{R}^2)$. It follows that with respect to this decomposition the Dirac operator on $\mathbf{R}^3$ is given by

$$\begin{pmatrix} i\nabla_{e_3} & -\nabla_{e_1} - i\nabla_{e_2} \\ \nabla_{e_1} - i\nabla_{e_2} & -i\nabla_{e_3} \end{pmatrix}$$

or we can rewrite this using the complex structure on $\mathbf{R}^2$ as

$$\begin{pmatrix} i\nabla_{e_3} & -2\frac{\partial}{\partial \bar{z}} \\ 2\frac{\partial}{\partial z} & -i\nabla_{e_3} \end{pmatrix}$$

where recall that the decomposition is $S_{\mathbf{C}}^-(\mathbf{R}^2) \oplus S_{\mathbf{C}}^+(\mathbf{R}^2)$.

The Index Theory of the Dirac Operator. Let X be a closed, oriented riemannian four-manifold. Let $\tilde{P}$ be a spin structure for X. Since $\partial\!\!\!/ : C^\infty(S_{\mathbf{C}}^+(\tilde{P})) \to C^\infty(S_{\mathbf{C}}^-(\tilde{P}))$ is an elliptic operator, it follows that the kernel is finite dimensional and that the image is a closed subspace of finite codimension. The index of $\partial\!\!\!/$ is by definition the complex dimension of the kernel minus the complex dimension of the cokernel. This index can be computed from the symbol in the following way. Consider the pullback $\pi^*(S_{\mathbf{C}}^\pm(\tilde{P}))$ to the cotangent bundle T^*X of $S_{\mathbf{C}}^\pm(\tilde{P})$. The symbol induces a bundle isomorphism between these bundles over the complement of the zero section of T^*X. In this way the symbol determines an element in the relative K-theory of $(T^*X, (T^*X - X))$. The Atiyah-Singer Index Theorem computes the index of the operator from this element in relative K-theory. In the case of the Dirac operator the index is $\hat{A}(X)$, the so-called A-hat genus of X. For a closed oriented four-manifold we have

$$\hat{A}(X) = \int_X -\frac{p_1(X)}{24} = -\frac{\sigma(X)}{8}$$

where $\sigma(X)$ is the signature of X. In the case of a *Spin^c*-structure $\tilde{P}$ on a four-manifold, the bundles in question change and the index formula gives

$$\frac{c_1(\det(\tilde{P}))^2 - \sigma(X)}{8}$$

as the index of the Dirac operator in this context.

Notice a couple of consequences of this theory: the $\hat{A}$-genus of a spin manifold is an integer. Also, the index of a spin four-manifold is divisible by 8 (actually, it is divisible by 16).

3.4   The Case of Complex Manifolds

Spin^c **structures for almost complex manifolds.**   Let V be a finite dimensional real inner product space with a complex structure given by $J: V \to V$ with $J^2 = -1$. We assume that this complex structure is compatible with the inner product in the sense that J is an orthogonal transformation. Let us decompose $V \otimes_{\mathbf{R}} \mathbf{C}$ into the i and $-i$ subspaces for the complexification of the action of J. We denote these subspaces by $V_{\mathbf{C}}^{1,0} \oplus V_{\mathbf{C}}^{0,1}$ and we denote the projections onto them by $\pi^{1,0}$ and $\pi^{0,1}$ respectively. Notice that $\pi^{1,0}$ is complex linear. This decomposition induces a bigrading of the complex exterior algebra $\Lambda_{\mathbf{C}}^* (V \otimes \mathbf{C})$. We denote by $\Lambda_{\mathbf{C}}^* V$ the subalgebra of the exterior powers of $V_{\mathbf{C}}^{1,0}$.

We define an action of V on $\Lambda_{\mathbf{C}}^* V$ by

$$v \cdot (\alpha^1 \wedge \cdots \wedge \alpha^t) = \sqrt{2} \left(\pi^{1,0}(v) \wedge \alpha^1 \wedge \cdots \wedge \alpha^t - \pi^{1,0}(v) \angle (\alpha^1 \wedge \cdots \wedge \alpha^t) \right)$$

where

$$\pi^{1,0}(v) \angle (\alpha^1 \wedge \cdots \wedge \alpha^t) = \sum_{i=1}^{t} (-1)^{i-1} \langle \alpha^i, \pi^{1,0}(v) \rangle \alpha^1 \wedge \cdots \wedge \hat{\alpha}^i \wedge \cdots \wedge \alpha^t$$

and the inner product is the hermitian inner product on $V \otimes \mathbf{C}$ (complex linear in the first factor) extending the given inner product on V. It is clear that this action is by complex linear endomorphisms of $\Lambda_{\mathbf{C}}^* V$, though it is not complex linear in V.

Lemma 3.4.1

$$v \cdot (v \cdot (\alpha^1 \wedge \cdots \wedge \alpha^t)) = -|v|^2 \alpha^1 \wedge \cdots \wedge \alpha^t.$$

Proof.   Clearly, both wedging with $\pi^{1,0}(v)$ and contracting with $\pi^{1,0}(v)$ are operations whose square is trivial. Thus, $v \cdot (v \cdot (\alpha^1 \wedge \cdots \wedge \alpha^k))$ is equal to

$$2 \left(-\pi^{1,0}(v) \angle (\pi^{1,0}(v) \wedge \alpha^1 \wedge \cdots \wedge \alpha^t) - \pi^{1,0}(v) \wedge (\pi^{1,0}(v) \angle (\alpha^1 \wedge \cdots \wedge \alpha^t)) \right)$$

which in turn is equal to

$$-2 \langle \pi^{1,0}(v), (\pi^{1,0}(v)) \rangle \alpha^1 \wedge \cdots \wedge \alpha^t.$$

Since $v \in V$, it is clear that

$$\langle \pi^{1,0}(v), \pi^{1,0}(v) \rangle = |\pi^{1,0}(v)|^2 = \frac{|v|^2}{2},$$

this completes the proof. $\square$

Corollary 3.4.2 *The action of V on $\Lambda^*_{\mathbf{C}}V$ extends to an action of $Cl(V)\otimes$ $\mathbf{C}$ by complex linear endomorphisms.*

Proof. We have seen that the relations defining the Clifford algebra hold for the given action. Since the action is complex linear, it determines an action of the complex Clifford algebra. □

Lemma 3.4.3 *The representation given in the previous corollary is, up to isomorphism, the spin representation.*

Proof. To prove this we need only see that this representation is irreducible. Since the dimension of $\Lambda^*_{\mathbf{C}}V$ is equal to that of the irreducible representation of $Cl(V) \otimes \mathbf{C}$ and since it is clear that this representation is non-trivial, the result is immediate. □

Next, we construct an embedding ρ of the unitary group $U(V)$ into $Spin^c(V)$. For any element $A \in U(V)$ there is an unitary frame $e_1, \cdots, e_n$ for V considered as a complex vector space in which A is diagonal; say $A(e_k) = e^{i\theta_k}e_k$, for $1 \le k \le n$. We associate to A the element in the complexification of the Clifford algebra of V given by

$$e^{\frac{i\theta_k}{2}} \prod_{k=1}^{n}(\cos(\frac{\theta_k}{2}) + \sin(\frac{\theta_k}{2})e_k J e_k)$$

where $\theta = \sum_{k=1}^{n} \theta_k$. It is clear that the given elements determine a continuous map from $U(V)$ to $Spin^c(V)$ sending the identity of $U(V)$ to the identity of $Spin^c(V)$. A direct computation shows that the projection of this mapping to $SO(V) \times S^1$ sends A to $(\iota(A), \det(A))$ where ι is the usual inclusion of $U(V) \subset SO(V)$. It follows that ρ is the unique lifting of $(\iota, \det)$ to a group homomorphism from $U(V)$ to $Spin^c(V)$.

Lemma 3.4.4 *Under the embedding ρ constructed above, the action of $U(V)$ on the spin module $\Lambda^*_{\mathbf{C}}V$ is the natural action of $U(V)$ on $\Lambda^*_{\mathbf{C}}(V)$.*

Proof. This is a direct computation. □

Corollary 3.4.5 *Let X be a compact riemannian manifold of dimension $2n$ with an almost complex structure $J: TX \to TX$ compatible with the metric. Then there is a natural $Spin^c$ structure $\tilde{P}_X$ for X whose determinant* line bundle is isomorphic to K_X^{-1}, the inverse of the canonical line bundle of $(0, n)$-forms for the almost complex structure. The complex spin bundle for this $Spin^c$ structure is isomorphic to the complex exterior algebra of the complex tangent bundle $\Lambda^* T_{\mathbb{C}} X$ where the action of the Clifford algebra is the one given in Corollary 3.4.2. Under this identification $S_{\mathbb{C}}^+(\tilde{P}_X)$ is $\Lambda^{2*} T_{\mathbb{C}} X$ and $S_{\mathbb{C}}^-(\tilde{P}_X)$ is $\Lambda^{2*+1} T_{\mathbb{C}} X$.

Proof. The complex structure and riemannian metric determine a reduction of the frame bundle of the tangent bundle to $U(n)$. That is to say we have a principal $U(n)$ bundle $Q_{U(n)} \to X$ and an isomorphism $Q_{U(n)} \times_{U(n)} SO(2n)$ with the orthogonal frame bundle $P_{SO(2n)}$ of X. We form

$$\tilde{P}_X = Q_{U(n)} \times_{U(n)} Spin^c(2n)$$

using the embedding $\rho: U(n) \to Spin^c(n)$ constructed above. Clearly,

$$\tilde{P}_X / S^1 = Q_{U(n)} \times_{U(n)} SO(2n)$$

where the representation $U(n) \to SO(2n)$ is the quotient of ρ. By construction this quotient is the natural embedding ι. This proves that $\tilde{P}_X$ is a $Spin^c$ structure on X. Its determinant line bundle of this $Spin^c$ structure is $\tilde{P}_X / Spin(2n) = Q_{U(n)} \times_{U(n)} S^1$ via the determinant map $U(n) \to S^1$. That is to say the determinant line bundle of $\tilde{P}_X$ is the determinant line bundle of $Q_{U(n)}$. This is the inverse line bundle to the determinant line bundle of $Q_{U(n)}^*$, the unitary frame bundle for the cotangent bundle. The determinant line bundle of the cotangent bundle is of course the canonical line bundle of $(0, n)$-forms. The fact that the Clifford multiplication on $S_{\mathbb{C}}(\tilde{P}_X)$ is as claimed is immediate from Corollary 3.4.2. Lastly, let us find the plus and minus one eigenspaces for multiplication by $\omega_{\mathbb{C}} = (i)^n e_1 \cdots e_{2n}$. We choose the coordinates so that $J e_{2k-1} = e_{2k}$. We denote by $d\bar{z}_j = e_{2j-1} + i e_{2j}$. Then Clifford multiplication by $\omega_{\mathbb{C}}$ is the same as Clifford multiplication by the element

$$i^n \frac{d\bar{z}_1}{2} \frac{i d\bar{z}_1}{2} \cdots \frac{i d\bar{z}_n}{2}.$$

Let us compute $\omega_{\mathbb{C}} \cdot (d\bar{z}_{j_1} \wedge \cdots \wedge d\bar{z}_{j_q})$. If $\ell \notin \{j_1, \ldots, j_q\}$, then

$$\frac{d\bar{z}_\ell}{2} \frac{i d\bar{z}_\ell}{2} \cdot (d\bar{z}_{j_1} \wedge \cdots \wedge d\bar{z}_{j_q}) = (-i)(d\bar{z}_{j_1} \wedge \cdots \wedge d\bar{z}_{j_q}),$$

whereas if $\ell \in \{j_1, \ldots, j_q\}$, then

$$\frac{d\bar{z}_\ell}{2} \frac{i d\bar{z}_\ell}{2} \cdot (d\bar{z}_{j_1} \wedge \cdots \wedge d\bar{z}_{j_q}) = i(d\bar{z}_{j_1} \wedge \cdots \wedge d\bar{z}_{j_q}).$$

Thus, we see that

$$\omega_{\mathbf{C}} \cdot (d\bar{z}_{j_1} \wedge \cdots \wedge d\bar{z}_{j_q}) = i^n(-i)^{n-q}i^q(d\bar{z}_{j_1} \wedge \cdots \wedge d\bar{z}_{j_q})$$
$$= (-1)^q(d\bar{z}_{j_1} \wedge \cdots \wedge d\bar{z}_{j_q}).$$

This completes the proof that the even degree forms are the $+1$ eigenspace for $\omega_{\mathbf{C}}$ and the odd degree forms are the -1 eigenspace. $\square$

We call the *Spin^c* structure given in the last corollary the *Spin^c structure determined by the almost complex structure*. We also say that it is the *Spin^c* structure of the almost complex structure.

Corollary 3.4.6 *The complex spin bundle of an almost complex manifold X with a compatible riemannian metric associated is identified with direct sum over all q of the exterior algebra bundle of complex-valued $(0,q)$-forms. Furthermore, $S^+_{\mathbf{C}}(\widetilde{P}_X)$ is identified with the bundle of $(0,2*)$-forms and $S^-_{\mathbf{C}}(\widetilde{P}_X)$ is identified with the bundle of $(0, 2*+1)$-forms. Lastly, Clifford multiplication by a vector field v on a $(0,q)$-form μ is given by*

$$v \cdot \mu = \sqrt{2}\left(\pi^{0,1}(v^*) \wedge \mu - \pi^{0,1}(v^*) \angle \mu\right)$$

where v^* is the dual one-form to v, $\pi^{0,1}$ denotes projection onto $\Lambda^{0,1}T^*X$ and $\angle$ is the contraction operator.

Proof. For a complex vector space V with a compatible inner product the duality map $V \to V^*$ given by $v \mapsto v^* = \langle \cdot, v \rangle$ induces by extension of scalars a complex linear map $V \otimes_{\mathbf{R}} \mathbf{C} \to \mathrm{Hom}_{\mathbf{R}}(V, \mathbf{C})$. This map induces a complex linear isomorphism $V^{1,0} \to (V^*)^{0,1}$. Taking the exterior powers of this isomorphism we see that $\Lambda^*_{\mathbf{C}}V$ is complex linearly isomorphic to $\Lambda^{0,*}_{\mathbf{C}}(V^*)$. Globalizing this isomorphism we see that the bundle of complex exterior powers of the tangent bundle of an almost complex manifold is complex linearly isomorphic to the bundle of $(0,q)$-forms on the manifold. Everything else is immediate from Corollary 3.4.5. $\square$

Remark 3.4.7 Notice that we can view differential forms $\Omega^i(X; \mathbf{R})$ as acting on $S_{\mathbf{C}}(\widetilde{P})$ by Clifford multiplication. We use the metric on X to identify $\Omega^i(X; \mathbf{R})$ with the space of sections of $\Lambda^i TX$ which is then identified with a subspace of the Clifford algebra $Cl(X)$. The action of i-forms is given as follows. Suppose that we have $\mu = a^1 \wedge \cdots \wedge a^i$ with the a^i being orthonormal at each point. Then Clifford multiplication by μ is the composition of Clifford multiplication by the α^j and Clifford multiplication by α^j is given by

$$a^j \cdot \nu = \sqrt{2} \left(\pi^{0,1}(a^j) \wedge \nu - \pi^{0,1}(a^j) \angle \nu \right).$$

The Dirac Operator on a Kähler Manifold. Now we assume that the almost complex structure is integrable and in fact is associated to a complex manifold X with a Kähler metric. This means that the manifold has a riemannian metric in which the J-operator of the complex structure is orthogonal, but also that at each point of the manifold there is a local holomorphic coordinate system in which the riemannian metric is standard to second order. We shall compute the Dirac operator at a point x using such a coordinate system. We take the $Spin^c$ structure induced from the complex structure. The determinant line bundle is identified with K_X^{-1}, the inverse of the canonical bundle. Of course, the complex structure on X determines a holomorphic structure (or equivalently a $(0,1)$-connection on K_X^{-1}). The metric on X determines a hermitian metric on K_X^{-1}. There is a unique hermitian connection A on K_X^{-1} which is compatible with the holomorphic structure in the sense that the $(0,1)$ part of the connection is the holomorphic $(0,1)$ connection. As we have just seen, the complex spin bundle of the $Spin^c$ structure is identified with the bundle of $(0,q)$-forms. Let σ be a $(0,k)$-form. Let $(z_1, \ldots, z_n)$ be holomorphic coordinates near a point p, so that setting $x_{2i-1} = \text{Re}(z_i)$ and $x_{2i} = \text{Im}(z_i)$ the riemannian metric is standard to second order at the point p. We let e_i be the unit tangent vector at p in the x_i-direction.

By definition for any section σ of the complex spin bundle we have

$$\partial\!\!\!/_A(\sigma)(p) = \sum_{i=1}^{2n} e_i \nabla_{e_i}(\sigma)(p).$$

Since the metric is standard to second order at the point, the connection A is product connection at the point and $\nabla_{e_i} = \partial_i$. Thus in the local trivialization,

$$\partial\!\!\!/_A(\sigma)(p) = \sum_{i=1}^{2n} e_i \partial_i(\sigma)(p)$$

which in turn is given by

$$\sqrt{2} \sum_{i=1}^{2n} \pi^{0,1}(dx_i) \wedge \partial_i(\sigma)(p) - \pi^{0,1}(dx_i) \angle \partial_i(\sigma)(p).$$

Of course $\pi^{0,1}(dx_{2k-1}) = d\bar{z}_k/2$ and $\pi^{0,1}(dx_{2k}) = id\bar{z}_k/2$. Thus, we have

$$\partial\!\!\!/_A(\sigma)(p) = \sqrt{2}\sum_{k=1}^{n} d\bar{z}_k \wedge \frac{1}{2}\left(\partial_{2k-1}(\sigma)(p) + i\partial_{2k}(\sigma)(p)\right)$$
$$-\sqrt{2}\sum_{i=1}^{2n} \pi^{0,1}(dx_i)\angle\partial_i(\sigma)(p)$$

which is

$$\sqrt{2}\sum_{k=1}^{n} d\bar{z}_k \wedge \frac{\partial}{\partial\bar{z}_k}(\sigma) - \sqrt{2}\sum_{i=1}^{2n} \pi^{0,1}(dx_i)\angle\partial_i(\sigma).$$

Since contraction is complex anti-linear in the first variable we can rewrite this last sum as

$$\partial\!\!\!/_A(\sigma)(p) = \sqrt{2}\sum_{k=1}^{n} d\bar{z}_k \wedge \frac{\partial}{\partial\bar{z}_k}(\sigma)(p)$$
$$-\sqrt{2}\sum_{k=1}^{n} (d\bar{z}_k)\angle\frac{1}{2}\left(\partial_{2k-1}(\sigma)(p) - i\partial_{2k}(\sigma)(p)\right).$$

This is the same as

$$\partial\!\!\!/_A(\sigma)(p) = \sqrt{2}\left(\bar{\partial}(\sigma)(p) - \sum_{k=1}^{n} d\bar{z}_k \angle \frac{\partial}{\partial z_k}(\sigma)(p)\right).$$

This last expression can be rewritten as

$$\partial\!\!\!/_A(\sigma)(p) = \sqrt{2}\left(\bar{\partial}(\sigma)(p) + \bar{\partial}^*(\sigma)(p)\right).$$

Chapter 4

The Seiberg-Witten Moduli Space

4.1 The Equations

We are now in a position to write down the Seiberg-Witten Equations for a closed, oriented riemannian four-manifold X. The equations require a $Spin^c$ structure $\widetilde{P}$ for the orthogonal frame bundle $P \to X$ of the tangent bundle of X. We denote by $\mathcal{L}$ the determinant line bundle of $\widetilde{P}$. From now on we denote the complex plus and minus spin bundles associated to $\widetilde{P}$ by $S^{\pm}(\widetilde{P})$. Recall that $S^{\pm}(\widetilde{P})$ has a hermitian inner product and hence so does $\mathcal{L}$ which is identified with the determinant line bundle of both $S^{\pm}(\widetilde{P})$. The Seiberg-Witten equations are equations for a spinor field $\psi \in C^{\infty}(S^{+}(\widetilde{P}))$ and a unitary connection A on $\mathcal{L}$. The equations are:

$$
\begin{aligned}
F_A^+ &= q(\psi) = \psi \otimes \psi^* - \frac{|\psi|^2}{2}\mathrm{Id} \\
\partial\!\!\!/_A(\psi) &= 0.
\end{aligned}
$$

Explanation of the Equations. The second equation needs little explanation: As described in the previous chapter, $\partial\!\!\!/_A$ is the Dirac operator associated to the Levi-Cívitá connection on the frame bundle of the tangent bundle and the connection A on the determinant line bundle of the $Spin^c$ structure. As to the first equation, since $S^{+}(\widetilde{P})$ has a hermitian metric, we can identify this complex vector bundle with its dual via an anti-complex isomorphism. The image of ψ under this isomorphism to the dual bundle

55

is denoted by ψ^*. Thus, we have

$$\psi \otimes \psi^* \in S^+(\widetilde{P}) \otimes S^+(\widetilde{P})^* = \text{End}_{\mathbf{C}}\left(S^+(\widetilde{P})\right).$$

(If we choose a basis for $S^+(\widetilde{P})$ at a point and then write

$$\psi = \begin{pmatrix} \psi_1 \\ \psi_2 \end{pmatrix}$$

with respect to this basis, then

$$\psi \otimes \psi^* = \begin{pmatrix} |\psi_1|^2 & \psi_1\overline{\psi_2} \\ \psi_2\overline{\psi_1} & |\psi_2|^2 \end{pmatrix}$$

with respect to the given basis.) We know that Clifford multiplication induces an isomorphism

$$(Cl_0(P) \otimes \mathbf{C})^+ \cong \text{End}_{\mathbf{C}}(S^+(\widetilde{P})).$$

We have also seen that

$$(Cl_0(P) \otimes \mathbf{C})^+ = \mathbf{C}\left(\frac{1 + \omega_{\mathbf{C}}}{2}\right) \oplus (\Lambda_+^2(TX) \otimes \mathbf{C}).$$

Under this isomorphism, $(1 + \omega_{\mathbf{C}})/2$ acts as the identity, and the traceless endomorphisms of $S^+(\widetilde{P})$ correspond to $\Lambda_+^2(TX) \otimes \mathbf{C}$. Clearly, the trace of $\psi \otimes \psi^*$ is $|\psi|^2$, so that

$$q(\psi) = \psi \otimes \psi^* - \frac{|\psi|^2}{2}\text{Id}$$

is traceless and hence is identified with a section of $\Lambda_+^2(TX) \otimes \mathbf{C}$. Using the metric to identity TX and T^*X (and extending complex linearly) $q(\psi)$ becomes complex-valued a self-dual two-form. The first equation simply says that this self-dual two-form is the self-dual part of the curvature.

There is one more lemma which needs to be established.

Lemma 4.1.1 *Under the above isomorphisms $q(\psi)$ is a purely imaginary self-dual two-form.*

Proof.   Recall that the real structure $Cl_0^+(\mathbf{R}^4) \subset (Cl_0(\mathbf{R}^4) \otimes \mathbf{C})^+$ is isomorphic to $\mathbf{H} \subset \mathbf{C}[2]$. That is to say the real Clifford algebra at a point consists of all two-by two complex matrices of the form

$$\begin{pmatrix} \alpha & -\overline{\beta} \\ \beta & \overline{\alpha} \end{pmatrix}.$$

Thus, an element in $\lambda \in \text{End}_{\mathbf{C}}(S^+(\widetilde{P}))$ is is in the real Clifford algebra if it satisfies

$$\text{Trace}(\lambda) \in \mathbf{R} \text{ and}$$

$$\lambda^* + \lambda = \text{Trace}(\lambda)\text{Id}.$$

It is clear from the construction that the trace of $q(\psi) = 0$. Thus, to see that $q(\psi)$ is the automorphism given by a totally imaginary self-dual two-form we need only see that

$$(iq(\psi))^* + iq(\psi) = 0.$$

But the dual of an element is simply the conjugate transpose. Thus, we need to establish that

$$-i\overline{q(\psi)}^{tr} + iq(\psi) = 0$$

or that

$$q(\psi) = \overline{q(\psi)}.$$

This is clear since it is obvious that such a relationship holds for $\psi \otimes \psi^*$ and for all real multiples of the identity. $\qquad\square$

4.2 Space of Configurations

In order to use these equations to produce a moduli space, out of which we will construct the Seiberg-Witten invariant of the *Spin^c* structure, we need to put these equations in a nonlinear elliptic framework. Let us begin by defining the space of configurations, the space on which the equations define a function. It is the space of all pairs (A, ψ) where A is a $U(1)$-connection on the determinant line bundle $\mathcal{L}$ of $\widetilde{P}$ and ψ is a section of $S^+(\widetilde{P})$. The naive way to do this would be to take C^∞-connections and sections. For technical reasons it is more convenient to work with Banach spaces or Hilbert spaces. Thus, the easiest thing to do, though by no means the only way to proceed, is to define $\mathcal{C}(\widetilde{P})$ to be the space

$$\mathcal{C}(\widetilde{P}) = \mathcal{A}_{L_2^2}(\mathcal{L}) \times L_2^2\left(S^+(\widetilde{P})\right)$$

where $\mathcal{A}_{L_2^2}(\mathcal{L})$ is the space of unitary L_2^2-connections on $\mathcal{L}$ and where for any vector bundle V over X the notation $L_k^2(V)$ means the space of L_k^2-sections of V. In the end we can work with any stronger norm; the moduli space is always the same and consists of C^∞ objects, up to gauge equivalence. The tangent space at any point to $\mathcal{C}(\widetilde{P})$ is naturally identified with

$$L_2^2\left((T^*X \otimes i\mathbf{R}) \oplus S^+(\widetilde{P})\right).$$

This space has its natural L^2-inner product.

We define the Seiberg-Witten function

$$F : \mathcal{C}(\widetilde{P}) \to L_1^2\left((\Lambda_+^2 T^* X \otimes i\mathbf{R}) \oplus S^-(\widetilde{P})\right)$$

by

$$F(A, \psi) = \left(F_A^+ - q(\psi), \partial\!\!\!/_A(\psi)\right).$$

In this context, the Seiberg-Witten equations are simply the equation

$$F(A, \psi) = 0.$$

Lemma 4.2.1 *The Seiberg-Witten functional F is a smooth mapping and its differential at (A, ψ) is given by the linear map*

$$DF_{(A,\psi)} = \begin{pmatrix} P_+ d & -Dq_\psi \\ \cdot\frac{1}{2}\psi & \partial\!\!\!/_A \end{pmatrix}$$

where $\cdot\frac{1}{2}\psi$ represents the Clifford multiplication map sending a one-form α to the element $\frac{1}{2}\alpha \cdot \psi \in L_1^2\left(S^-(\widetilde{P})\right)$. Furthermore,

$$Dq_\psi(\eta) = \psi \otimes \eta^* + \eta \otimes \psi^* - \frac{\langle \eta, \psi \rangle + \overline{\langle \eta, \psi \rangle}}{2} \mathrm{Id}.$$

Again this element is a traceless, self-adjoint automorphism and hence is identified via Clifford multiplication with a purely imaginary self-dual two-form.

Proof. All the computations of the derivatives are straightforward. Notice that F is an affine mapping plus the quadratic mapping $q(\psi)$. The affine map is continuous and hence smooth. It is obvious from the Sobolev multiplication theorem that $q(\psi)$ is a smooth map. □

4.3 Group of Changes of Gauge

We must also consider the group of changes of gauge or bundle automorphisms. We choose these to be the automorphisms of the principal $Spin^c$ bundle $\widetilde{P}$ which cover the identity on the frame bundle of the tangent bundle. Such an automorphism is given by a map from the manifold X to the center S^1 of $Spin^c(4)$. As is usual in these contexts, one needs to control one more derivative on the bundle automorphism in order to have an action. Thus, we take L^2_3-maps. (Note the consideration that dictates which norms we work with is the fact that the group of gauge automorphisms must be continuous bundle automorphisms, and hence at least L^2_3.) We denote the space of such mappings with the L^2_3-topology as $\mathcal{G}(\widetilde{P})$. This is a Hilbert manifold whose tangent space at the identity is the L^2_3-functions on X with values in $i\mathbf{R} \subset \mathbf{C}$.

Lemma 4.3.1 *Pointwise multiplication makes $\mathcal{G}(\widetilde{P})$ into an infinite dimensional abelian Lie group. Its Lie algebra is $L^2_3(X; i\mathbf{R})$ (the space of L^2_3-sections of the trivial bundle $X \times i\mathbf{R}$ over X) with the trivial bracket.*

Notice that this Lie algebra has a natural L^2-inner product.

Proof. The only thing to establish is that multiplication and inverses are smooth mappings of Hilbert manifolds. This is clear for inverse; let us consider multiplication. The point is that there is a continuous Sobolev multiplication

$$L^2_3(X) \otimes L^2_3(X) \to L^2_3(X).$$

From this the result is immediate. □

4.4 The Action

Next we show that there is a natural action of the group of changes of gauge on the space of configurations. If $\sigma \in \mathcal{G}(\widetilde{P})$ then there are induced bundle maps $\det \sigma$ on $\mathcal{L}$ and $S^{\pm}(\sigma)$ on $S^{\pm}(\widetilde{P})$. Notice that if we view $\det \sigma$ as a function from X to S^1, then it is simply the image of the function σ under the squaring map from S^1 to S^1. The action is given by

$$(A, \psi) \cdot \sigma = ((\det \sigma)^* A, S^+(\sigma^{-1})(\psi)).$$

Lemma 4.4.1 *The above formula defines a smooth right action*

$$\mathcal{C}(\widetilde{P}) \times \mathcal{G}(\widetilde{P}) \to \mathcal{C}(\widetilde{P}).$$

The transformation law from the function F is:

$$F((A, \psi) \cdot \sigma) = F(A, \psi) \cdot \sigma$$

where the action of σ on $L_1^2\left((\Lambda_+^2 T^* X \otimes i\mathbf{R}) \oplus S^-(\widetilde{P})\right)$ is trivial on the first factor and is given by $S^-(\sigma^{-1})$ on the second factor.

Proof. The fact that the action is smooth is easily established from the Sobolev multiplication theorems.

In order to prove the transformation law for F we need the following.

Lemma 4.4.2 *Let $\sigma \in \mathcal{G}$ and let $\psi \in S^+(\widetilde{P})$. Then*

$$\partial\!\!\!/_{(\det \sigma)^* A}\left(S^+(\sigma^{-1})(\psi)\right) = S^-(\sigma^{-1})(\partial\!\!\!/_A(\psi)).$$

Proof. First notice that if ∇' is the covariant derivative on $S^\pm(\widetilde{P})$ determined by $(\det \sigma)^* A$ and the Levi-Cívitá connection and if ∇ is the covariant derivative on $S^\pm(\widetilde{P})$ determined by A and the Levi-Cívitá connection, then

$$\nabla' = S^\pm(\sigma)^* \nabla.$$

It then follows that

$$\nabla'\left(S^\pm(\sigma^{-1})(\psi)\right) = S^\pm(\sigma^{-1})(\nabla(\psi)).$$

From this the lemma is direct since Clifford multiplication commutes with the automorphisms $S^\pm(\sigma)$. $\qquad\square$

The first lemma is now immediate from this result. In particular, it follows that the space of solutions to the Seiberg-Witten equations is invariant under the action of $\mathcal{G}(\widetilde{P})$. $\qquad\square$

4.5   The Quotient Space

Now we need to pass to the quotient space of the action of $\mathcal{G}(\widetilde{P})$ on $\mathcal{C}(\widetilde{P})$. It is not a *priori* obvious that there is a reasonable quotient. Before getting into these questions, let us describe the various type of stabilizers that appear in this action. Throughout this section we implicitly assume that the base manifold X is connected.

Lemma 4.5.1 *The stabilizer in $\mathcal{G}(\widetilde{P})$ of an element $(A, \psi) \in \mathcal{C}(\widetilde{P})$ is trivial unless $\psi = 0$ in which case the stabilizer is the group consisting of the constant maps of X to S^1, a group naturally identified with S^1.*

Definition 4.5.2 We say that an element (A, ψ) is *irreducible* if $\psi \neq 0$, otherwise it is *reducible*. We denote by $\mathcal{C}^*(\widetilde{P})$ the open subset of irreducible configurations.

Proof. Since the manifold X is connected, the stabilizer of any connection A is exactly the group of constant maps from X to S^1. This subgroup acts freely on ψ unless ψ is identically zero, in which case it acts trivially on ψ. $\qquad\square$

The basic convergence result. Now let us turn to the more technical results on the nature of the action. Here is the basic lemma.

Lemma 4.5.3 *Suppose that (a_n, ψ_n) and (b_n, μ_n) are sequences in $\mathcal{C}(\widetilde{P})$ converging to (a, ψ) and (b, μ) respectively. Suppose also that for each n we have $\sigma_n \in \mathcal{G}(\widetilde{P})$ with*

$$(a_n, \psi_n) \cdot \sigma_n = (b_n, \mu_n).$$

Then there is a subsequence of the σ_n which converges to an element $\sigma \in \mathcal{G}(\widetilde{P})$. Furthermore, we have

$$(a, \psi) \cdot \sigma = (b, \mu).$$

Proof. Since the σ_n are all functions with values in the circle. It is clear that we have a bound on $\|\sigma_n\|_{L^6}$ independent of n. Let $\tau_n = \det(\sigma_n)$. Clearly, we also have a bound on $\|\tau_n\|_{L^6}$ independent of n. The fact that $\tau_n^* a_n = b_n$ means that
$$d\tau_n = \tau_n(b_n - a_n).$$
Since the sequences $\{a_n\}$ and $\{b_n\}$ converge to L_2^2-connections a and b respectively, clearly $\|a_n\|_{L_2^2}$ and $\|b_n\|_{L_2^2}$ are bounded independent of n. Using the Sobolev multiplication $L^6 \otimes L_2^2 \to L^5$, we see that $\|d\tau_n\|_{L^5}$ is bounded independent of n. This gives a bound on $\|\tau_n\|_{L_1^5}$ independent of n. Continuing in this bootstrap manner we have bounds for $\|\tau_n\|_{L_2^4}$ and $\|\tau_n\|_{L_3^3}$ which are independent of n. Now by the compactness of the embedding $L_3^3 \subset L_2^3$ we see that it is possible to extract a subsequence of

the τ_n which converges in L_2^3 to an element τ. Clearly, $d\tau = \tau(b-a)$ so that once again we see by Sobolev multiplication that $d\tau \in L_2^2$ and that $d\tau_n$ converges to $d\tau$ in L_2^2. It follows that $\tau \in L_3^3$ and that the τ_n converge to τ in L_3^3. Thus, we have established that after extracting a subsequence we can arrange that $(\sigma_n)^2$ converge in L_3^2 to an element τ. Passing to a further subsequence, we find a subsequence of the σ_n which converge in L_3^2 to an element σ. It is obvious that $(a, \psi) \cdot \sigma = (b, \mu)$. $\square$

Corollary 4.5.4 *The quotient space $\mathcal{B}(\widetilde{P}) = \mathcal{C}(\widetilde{P})/\mathcal{G}(\widetilde{P})$ is a Hausdorff space.*

Proof.   Since the spaces in question are first countable, if the quotient is not Hausdorff, then there is a sequence (a_n, ψ_n) in $\mathcal{C}(\widetilde{P})$ and a sequence $\sigma_n \in \mathcal{G}(\widetilde{P})$ so that (a_n, ψ_n) converges to say (a, ψ) and $(a_n, \psi_n) \cdot \sigma_n$ converges to say (b, μ) and furthermore, (a, ψ) and (b, μ) are not in the same orbit of the action. But this is impossible – after passing to a subsequence, we can assume that the σ_n converge to σ and $(a, \psi) \cdot \sigma = (b, \mu)$. $\square$

That is but the first result we require about the quotient space. We also need the slice theorem.

The Slice Theorem.

Lemma 4.5.5 *There are local slices for the action of $\mathcal{G}(\widetilde{P})$ on $\mathcal{C}(\widetilde{P})$.*

Remark 4.5.6 By *local slices* for the action, we mean that for each $x \in \mathcal{C}(\widetilde{P})$ there is an open neighborhood of the point and a smoothly embedded (closed) Hilbert submanifold S of this neighborhood invariant under the stabilizer of the x such that the natural map

$$S \times_{\mathrm{Stab}(x)} \mathcal{G}(\widetilde{P}) \to \mathcal{C}(\widetilde{P})$$

is a diffeomorphism onto a neighborhood of the orbit through x.

Proof.   First, we consider the differential of the action of $\mathcal{G}(\widetilde{P})$ at a point $(A, \psi) \in \mathcal{C}(\widetilde{P})$. This differential is given by the map

$$L_3^2(X; i\mathbf{R}) \to L_2^2\left((T^*X \otimes i\mathbf{R}) \oplus S^+(\widetilde{P})\right)$$

which is given by

$$f \mapsto (2df, -f \cdot \psi).$$

The linear version of the slice is simply the kernel of the adjoint of this mapping, adjoint with respect to the L^2-inner products on domain and range. That is to say, we consider the subspace $(A, \psi) + K$ where K is the kernel of the linear map

$$\Omega^1(X; i\mathbf{R})_{L^2_2} \oplus L^2_2(S^+(\widetilde{P})) \to \Omega^0(X; i\mathbf{R})_{L^2_1}$$

given by

$$(\omega, \mu) \mapsto 2d^*\omega + i\text{Im}(<\mu, \psi>_{L^2}).$$

Notice that the linear subspace K is invariant under the $\text{Stab}(A, \psi)$ action.

The slice for the action will $(A, \psi) + U$ where U is a sufficiently small $\text{Stab}(A, \psi)$ open neighborhood of the origin in K. Let us choose such a neighborhood U and consider the map

$$U \times_{\text{Stab}(A,\psi)} \mathcal{G}(\widetilde{P}) \to \mathcal{C}(\widetilde{P})$$

given by sending

$$[u, \sigma] \mapsto ((A, \psi) + u) \cdot \sigma.$$

This is a well-defined smooth mapping. Its differential at $[0, \text{Id}]$ is the map

$$\left(K \oplus L^2_3(X; i\mathbf{R})\right) /\text{stab}(A, \psi) \to L^2_2\left((T^*X \otimes i\mathbf{R}) \oplus S^+(\widetilde{P})\right)$$

where $\text{stab}(A, \psi)$ is the tangent space to $\text{Stab}(A, \psi)$ and the map is the inclusion on the first factor and $(2d, -(\cdot)\psi)$ on the second factor. Since K is the orthogonal complement to the image of the map $(2d, -(\cdot)\psi)$, and since $\text{stab}(A, \psi)$ is the kernel of this map, it follows that the above map is a linear isomorphism. Hence, by the inverse function theorem, if we choose U sufficiently small and if we let $V \subset \mathcal{G}(\widetilde{P})$ be a sufficiently small $\text{Stab}(A, \psi)$ invariant neighborhood of the identity in $\mathcal{G}(\widetilde{P})$, then the map

$$U \times_{\text{Stab}(\widetilde{P})} V \to \mathcal{C}(\widetilde{P})$$

is a diffeomorphism onto an open neighborhood of (A, ψ).

We need to show that, possibly after replacing U by a smaller open neighborhood, the map

$$U \times_{\text{Stab}(A,\psi)} \mathcal{G}(\widetilde{P}) \to \mathcal{C}(\widetilde{P})$$

is a diffeomorphism onto an open subset. Since, as we have just seen, the map is a local diffeomorphism, the only thing to show is that for U sufficiently small, the map is one-to-one. If there is no such smaller neighborhood U of 0 where the map is one-to-one, then we can find sequences

$\{a_n\}_{n=1}^{\infty}$ and $\{b_n\}_{n=1}^{\infty}$ of points in U converging to 0 and group elements $\{\sigma_n\}$ such that

$$((A, \psi) + a_n) \cdot \sigma_n = (A, \psi) + b_n$$

but with $[a_n, \sigma_n] \neq [b_n, \mathrm{Id}]$ in the fibered product. Lemma 4.5.3 allows us to extract a subsequence so that the σ_n converge to an element $\sigma \in \mathcal{G}(\widetilde{P})$. Clearly, $(A, \psi) \cdot \sigma = (A, \psi)$; that is to say $\sigma \in \mathrm{Stab}(A, \psi)$. This means that for all n sufficiently large $[a_n, \sigma_n]$ and $[b_n, \mathrm{Id}]$ are both contained in $U \times_{\mathrm{Stab}(A,\psi)} V$. By the fact that the restriction of the map to this subset is a diffeomorphism, we obtain a contradiction. This proves the diffeomorphism statement and completes the proof of the fact that slices for the action exist.

$\square$

Corollary 4.5.7 *The quotient space $\mathcal{B}(\widetilde{P}) = \mathcal{C}(\widetilde{P})/\mathcal{G}(\widetilde{P})$ is a Hausdorff space. The complement of the equivalence classes of reducible configurations forms an open subset, denoted $\mathcal{B}^*(\widetilde{P})$, which is a Hilbert manifold. Its tangent space at $[A, \psi]$ is identified with*

$$L_2^2 \left((T^*X \otimes i\mathbf{R}) \oplus S^+(\widetilde{P}) \right) /\mathrm{Image}(2d, -(\cdot)\psi).$$

If $[A, 0]$ is a reducible equivalence class, then a neighborhood of this point in the quotient space is homeomorphic to the quotient of

$$L_2^2 \left((T^*(X \otimes i\mathbf{R}) \oplus S^+(\widetilde{P})) \right)(\mathrm{Image}(2d), 0)$$

by the action of $S^1 = \mathrm{Stab}(A, 0)$. This action is linear and semi-free. The fixed points for the tangent space to the subspace of reducible equivalence classes. In particular, the link of the stratum of reducible equivalence classes in the quotient space is homotopy equivalent to $\mathbf{C}P^{\infty}$.

The homotopy type of the quotient is also easy to establish.

Lemma 4.5.8 *Let $\mathcal{B}^*(\widetilde{P})$ be the open subset of $\mathcal{B}(\widetilde{P})$ consisting of irreducible equivalence classes. Then $\mathcal{B}^*(\widetilde{P})$ is a classifying space for the group $(S^1)^X$. In particular, it is homotopy equivalent to a product of $\mathbf{C}P^{\infty}$ with an Eilenberg-MacLane space $K(H^1(X; \mathbf{Z}), 1)$.*

Proof. The space $\mathcal{C}(\widetilde{P})$ is an affine space and hence is contractible. The subset of reducible objects is an affine subspace of infinite codimension, so that the open subset $\mathcal{C}^*(\widetilde{P})$ of irreducible objects is also contractible. It follows that, since the action of $(S^1)^X$ on the complement is free and has local slices, the quotient space $\mathcal{B}^*(\widetilde{P})$ is a classifying space for $(S^1)^X$. This group is the product of discrete group $H^1(X; \mathbf{Z})$ with the component of the identity, which is the group of all maps $X \to S^1$ which are homotopically trivial. This later group deformation retracts onto the group of constant maps and hence is homotopy equivalent to S^1. The statements about the homotopy type of the classifying space are now immediate. $\qquad \square$

There is a universal S^1-bundle over $\mathcal{B}^*(\widetilde{P})$ whose first Chern class is the generator for $H^2(\mathcal{B}^*(\widetilde{P}); \mathbf{Z})$ corresponding to the $\mathbf{C}P^\infty$ factor. Fix a base point $x \in X$ and let $\mathcal{G}^0(\widetilde{P})$ be the subgroup of $\mathcal{G}(\widetilde{P})$ consisting of all changes of gauge which are trivial on the fiber over x. Clearly, $\mathcal{G}^0(\widetilde{P})$ is the kernel of the homomorphism $\mathcal{G}(\widetilde{P}) \to S^1$ given by evaluating on the fiber over x. Let $\mathcal{B}^0(\widetilde{P})$ be the quotient $\mathcal{C}^*(\widetilde{P})/\mathcal{G}^0(\widetilde{P})$. It is the total space of the required principal circle bundle over $\mathcal{B}^*(\widetilde{P})$. It is easy to see that its Chern class generates the second cohomology of the $\mathbf{C}P^\infty$ factor of this space. Furthermore, the restriction of the first Chern class of this principal S^1-bundle to the link of the reducible stratum is a generator of the second cohomology of the link (which recall is homotopy equivalent to $\mathbf{C}P^\infty$).

4.6 The Elliptic Complex

Let $(A, \psi) \in \mathcal{C}(\widetilde{P})$ be a solution to the Seiberg-Witten equations. There is a naturally associated elliptic complex which incorporates the linearization of the action of the group of gauge transformations and the linearization of the Seiberg-Witten equations. It is denoted $(\mathcal{E}(A, \psi))$ and is the complex

$$0 \longrightarrow L_3^2(X; i\mathbf{R}) \longrightarrow L_2^2\left((T^*X \otimes i\mathbf{R}) \oplus S^+(\widetilde{P})\right) \longrightarrow$$
$$\longrightarrow L_1^2\left((\Lambda_+^2 T^*X \otimes i\mathbf{R}) \oplus S^-(\widetilde{P})\right) \longrightarrow 0$$

where the first map is $(2d, -(\cdot)\psi)$ and the second is given by the matrix

$$\begin{pmatrix} P_+d & -Dq_\psi \\ \cdot\frac{1}{2}\psi & \partial\!\!\!/_A \end{pmatrix}.$$

We claim that this is a complex and that it is elliptic. We shall also compute its Euler characteristic.

Claim 4.6.1 *For any solution (A, ψ) to the Seiberg-Witten equations the above is a complex.*

Proof. We begin with $\varphi \in L_3^2(X; i\mathbf{R})$. Its image under the first map is $(2d\varphi, -\varphi\psi)$, and hence its image under the composition of the two maps is

$$(2P_+dd\varphi + Dq_\psi(\varphi\psi), d\varphi \cdot \psi - \partial\!\!\!/_A(\varphi\psi)) .$$

Of course, $2P_+dd\varphi = 0$ and $Dq_\psi(\varphi\psi)$ is equal to

$$\psi \otimes \overline{\varphi}\psi^* + \varphi\psi \otimes \psi^* - \mathrm{Re}(\langle \psi, \varphi\psi\rangle\mathrm{Id}.$$

Since φ is purely imaginary $\overline{\varphi} = -\varphi$, and hence the first two terms cancel each other. The last term is zero for the same reason. This shows that the first component of the composition is trivial.

Let us consider the second component. The Dirac operator acts like a differentiation over the functions so that

$$\partial\!\!\!/_A(\varphi\psi) = d\varphi \cdot \psi + \varphi\partial\!\!\!/_A(\psi).$$

From this it is clear that the second component of the composition is equal to $-\partial\!\!\!/_A(\psi)$ which is zero since (A, ψ) is a solution to the Seiberg-Witten equations. □

Suppose that $[A, \psi]$ is an irreducible solution to the Seiberg-Witten equations. Then the zeroth cohomology of the complex is trivial; the first cohomology of this complex is a finite dimensional linear subspace of $T\mathcal{B}^*(\widetilde{P})$ at the point $[A, \psi]$. This is the *Zariski tangent space* (or *formal tangent space*) to the moduli space at this point. The second cohomology of the complex is called the *obstruction space* at $[A, \psi]$. We see that the dimension of the Zariski tangent space minus the dimension of the obstruction space is equal to minus the Euler characteristic of the elliptic complex. We say that $\mathcal{M}(\widetilde{P})$ is *smooth* at an irreducible solution if the obstruction space is trivial. It follows from the implicit function theorem that if $\mathcal{M}(\widetilde{P})$ is smooth at an irreducible point $[A, \psi]$, then a neighborhood of $[A, \psi]$ in $\mathcal{M}(\widetilde{P})$ is a smooth submanifold of $\mathcal{B}^*(\widetilde{P})$ whose tangent space at $[A, \psi]$ is the Zariski tangent space at this point.

Now let us compute the Euler characteristic of this elliptic complex. First notice that we can deform this complex by a homotopy to get rid of the zeroth order terms. This homotopy will not change the symbol sequence or the Euler characteristic of the complex. The resulting deformed complex is simply the direct sum of two complexes:

$$0 \xrightarrow{} L_3^2(X; i\mathbf{R}) \xrightarrow{2d} L_2^2(T^*X \otimes i\mathbf{R}) \xrightarrow{P_+d} L_1^2\left(\Lambda_+^2 T^*X \otimes i\mathbf{R}\right)$$

$$\xrightarrow{} 0$$

and

$$0 \longrightarrow 0 \longrightarrow L_2^2\left(S^+(\widetilde{P})\right) \xrightarrow{\ \partial_A\ } L_1^2\left(S^-(\widetilde{P})\right) \longrightarrow 0.$$

Clearly, each of these complexes is elliptic, proving that the original one is also. The Euler characteristic of the first complex is $1 + b_2^+(X) - b_1(X)$ whereas the (complex) Euler characteristic of the second complex is equal to minus the index of the dirac operator which we have seen is equal to

$$(\operatorname{sign}(X) - c_1(\mathcal{L})^2)/8.$$

We are interested in the real Euler characteristic which is then given by

$$1 + b_2^+(X) - b_1(X) + \left(\operatorname{sign}(X) - c_1(\mathcal{L})^2\right)/4.$$

This simplifies to

$$\left(2\chi(X) + 3\operatorname{sign}(X) - c_1(\mathcal{L})^2\right)/4.$$

Corollary 4.6.2 *Let $[A, \psi] \in \mathcal{M}(\widetilde{P})$ be an irreducible point. Then the dimension of the Zariski tangent space to $\mathcal{M}(\widetilde{P})$ at $[A, \psi]$ minus the dimension of the obstruction space at $[A, \psi]$ is equal to*

$$d = \left(c_1(\mathcal{L})^2 - 2\chi(X) - 3\operatorname{sign}(X)\right)/4.$$

In particular, if $[A, \psi]$ is a smooth irreducible point of $\mathcal{M}(\widetilde{P})$, then near $[A, \psi]$ the space $\mathcal{M}(\widetilde{P})$ is a smooth manifold of dimension d.

Chapter 5

Curvature Identities and Bounds

The main result in this chapter is the compactness of the moduli space of solutions to the Seiberg-Witten equations. In fact, it is a very strong form of compactness: for any metric on X there are only finitely many non-empty moduli spaces and each of these is compact. (Actually, there is a hidden assumption here, we are dealing only with moduli spaces whose formal dimensions are non-negative. Of course, if we choose a generic perturbation of the equations, then the only non-empty moduli spaces will have formal dimensions which are non-negative.) This compactness result is a consequence of two phenomena. First, we have the general compactness criterion of Uhlenbeck, which applies in this context to show that L^2-bounds on the curvature of the connections suffice to imply L_1^2-bounds on the connections in a appropriate gauge. The other result, which holds here because of the special nature of the equations, but which does not hold in the case of ASD connections is an a *priori* pointwise bound on the spinor field and hence on the self-dual part of the curvature of any solution. These, together with the bound of c_1^2 coming from the assumption that formal dimension of the moduli space is non-negative, imply an L^2-bound on the curvature.

We begin with the curvature facts needed to establish the a *priori* bounds.

5.1   Curvature Identities

Suppose that ∇ is Levi-Cività connection on the tangent bundle of a riemannian manifold X. Let R be the curvature of this connection: thus R is a two-form with values in the skew-adjoint endomorphisms of the tangent bundle. Let $U, V \in T_x X$. We denote by $R(U, V) : T_x X \to T_x X$ the skew adjoint endomorphism obtained by evaluating the curvature form on the ordered pair of tangent vectors (U, V). Of course, since $R = \nabla \circ \nabla$ we have

$$R(U, V) = \nabla_U \circ \nabla_V - \nabla_V \circ \nabla_U$$

as endomorphisms of $T_x X$.

Lemma 5.1.1 *For any three tangent vectors U, V, W at a point $x \in X$ we have the identity in $T_x X$*

$$R(U, V)(W) + R(V, W)(U) + R(W, U)(V) = 0.$$

Proof. Recall that the torsion-free condition is then

$$\nabla_A(B) = \nabla_B(A)$$

provided that A and B are extended to vector fields in a neighborhood which commute. We extend U, V, W to vector fields on a neighborhood of $x \in X$ which have trivial brackets. Since the curvature is a tensor, the value of the expression in the statement of the lemma is independent of the extension of U, V, W. Of course, the curvature form is simply the two-form with values in the endomorphisms given by $\nabla \circ \nabla$ mapping sections of TX to two-forms with values in TX. Since the vector fields extending U, V, W have trivial brackets with each other, this means that

$$R(U, V)(W) = \nabla_U \circ \nabla_V(W) - \nabla_V \circ \nabla_U(W)$$

and similarly for the other permutations. Plugging all this in we have:

$$
\begin{aligned}
&R(U, V)(W) + R(V, W)(U) + R(W, U)(V) \\
&= \quad \nabla_U \circ \nabla_V(W) - \nabla_V \circ \nabla_U(W) + \nabla_V \circ \nabla_W(U) \\
&\quad -\nabla_W \circ \nabla_V(U) + \nabla_W \circ \nabla_U(V) - \nabla_U \circ \nabla_W(V).
\end{aligned}
$$

Using the torsion-free condition to reverse the last two tangent vectors in the third, fifth, and sixth terms yields the desired vanishing result.   □

Definition 5.1.2 Let ∇ be the covariant derivative of the riemannian connection on a riemannian manifold X. Let R be its curvature two-form. The *Ricci curvature* of X is the bilinear form on TX given by

$$\text{Ric}(V, W) = \langle -\sum_i R(e_i, V)(e_i), W \rangle$$

where $(e_1, \ldots, e_n)$ is an orthonormal basis for $T_x X$. Let $(x_1, \ldots, x_n)$ be a coordinate system for which the standard tangent vectors form an orthonormal system at a point x. Then the curvature form at this point is given by

$$R = \sum_{k,\ell} \left(R_{i,j}^{k,\ell} \right) dx_k \wedge dx_\ell.$$

Thus, in this notation the upper indices refer to the two-form indices and the lower indices are the matrix indices in the Lie algebra. This tensor is skew-symmetric in both the upper two indices and the lower two indices. The curvature two-form can also be written as

$$\sum_{i<j} \sum_{k<\ell} -R_{i,j}^{k,\ell} dx^k \wedge dx^\ell (e_i \wedge e_j) = \sum_{i<j} \sum_{k<\ell} R_{j,i}^{k,\ell} dx^k \wedge dx^\ell (e_i \wedge e_j)$$

where, as before, $e_i \wedge e_j$ denotes the infinitessimal generator of the rotation in the (i, j)-plane taking e_i toward e_j. The minus sign is present since $e_i \wedge e_j$ is represented by the matrix with 1 in the (j, i)-place and -1 in the (i, j)-place.

Thus, $\langle R(e_i, e_j)(e_i), e_k \rangle = -R_{i,k}^{i,j}$, and hence

$$\text{Ric}(e_j, e_k) = \sum_i R_{i,k}^{i,j}.$$

Lemma 5.1.3 *The Ricci curvature is well-defined and is a symmetric bilinear form.*

Proof. Let us consider two orthonormal bases $(e_1, \ldots, e_n)$ and $(e_1', \ldots, e_n')$ with $e_i' = \sum_j A_{i,j} e_j$ for some orthogonal matrix A. Then since R is linear in each of its variables we have

$$
\begin{aligned}
\sum_i \langle R(e_i', V)(e_i'), W \rangle &= \sum_i \sum_j A_{i,j} \langle R(e_j, V)(e_i'), W \rangle \\
&= \sum_i \sum_j \sum_k A_{i,j} A_{i,k} \langle R(e_j, V)(e_k), W \rangle \\
&= \sum_{j,k} \left(\sum_i A_{i,j} A_{i,k} \langle R(e_j, V)(e_k), W \rangle \right).
\end{aligned}
$$

Since A is an orthogonal matrix we have

$$\sum_i A_{i,j} A_{i,k} = \delta_{j,k}.$$

Thus,

$$\sum_i \langle R(e_i', V)(e_i'), W \rangle = \sum_{j,k} \delta_{j,k} \langle R(e_j, V)(e_k), W \rangle = \sum_j \langle R(e_j, V)(e_j), W \rangle.$$

This completes the proof of the fact that the Ricci curvature is well-defined.

Let us turn now to the fact that it is symmetric. First, notice that it is clear that $\langle R(U, V)(W), Y \rangle$ is skew symmetric in the first two variables since the curvature is a two-form. It is also skew symmetric in the last two variables since the values of the curvature form lie in the Lie algebra of the orthogonal group which is the skew symmetric endomorphisms of $\mathbf{R}^n$. We claim that it is symmetric under interchanging the first and second variables with the third and fourth variables. From this symmetry, the symmetry of the Ricci curvature is immediate.

To establish this symmetry we compute using Lemma 5.1.1 and the skew symmetry:

$$
\begin{aligned}
\langle R(U, V)(W), Y \rangle &= -\langle R(W, U)(V), Y \rangle - \langle R(V, W)(U), Y \rangle \\
&= \langle R(W, U)(Y), V \rangle + \langle R(V, W)(Y), U \rangle \\
&= -\langle R(Y, W)(U), V \rangle - \langle R(U, Y)(W), V \rangle \\
&\quad -\langle R(W, Y)(V), U \rangle - \langle R(Y, V)(W), U \rangle \\
&= 2\langle R(W, Y)(U), V \rangle + \langle R(U, Y)(V), W \rangle \\
&\quad +\langle R(Y, V)(U), W \rangle \\
&= 2\langle R(W, Y)(U), V \rangle - \langle R(V, U)(Y), W \rangle \\
&= 2\langle R(W, Y)(U), V \rangle - \langle R(U, V)(W), Y \rangle.
\end{aligned}
$$

From this the symmetry is clear.   □

There is a related curvature which is simply the trace of the Ricci curvature thought of as a symmetric endomorphism of TX.

Definition 5.1.4 The *scalar curvature* of a riemannian manifold X is defined to be a function

$$\kappa \colon X \to \mathbf{R}$$

given by

$$\kappa = \text{Trace}(\text{Ric}).$$

In terms of the curvature two-form we have

$$\kappa = \sum_{i,j} R_{i,j}^{i,j} = 2 \sum_{i<j} R_{i,j}^{i,j}.$$

Notice that if X is a riemann surface then the scalar curvature is equal to twice the Gaussian curvature.

Now we come to a Bochner-type formula which plays an important role in the theory.

Proposition 5.1.5 *Let X be a riemannian manifold and let $\widetilde{P} \to X$ be a Spin^c structure for X. Let A be a connection on $\det(\widetilde{P})$ and let $\partial\!\!\!/_A$ be the Dirac operator on $S(\widetilde{P})$ determined by the Levi-Cività connection on the orthogonal frame bundle and A on the determinant line bundle. Then for any section ψ of $S(\widetilde{P})$ we have*

$$\partial\!\!\!/_A \circ \partial\!\!\!/_A(\psi) = \nabla_A^* \nabla_A(\psi) + \frac{\kappa}{4}\psi + \frac{F_A}{2} \cdot \psi \qquad (5.1)$$

where $F_A \in \Omega^2(X; i\mathbf{R})$ is the curvature of A and the last term is Clifford multiplication by the two-form.

Proof. Let R be the two-form with values in the endomorphisms of TX which is the curvature form for the Levi-Cività connection and let F_A be the purely imaginary two-form which is the curvature form of A. Then the curvature form F for the covariant derivative on the fibered product of the frame bundle of the tangent bundle and the determinant line bundle is

$$F = R + F_A.$$

Writing the curvature on the frame bundle R as $-\sum_{i<j} R_{i,j} e_i \wedge e_j$ as in Chapter 3 we see that the action of F on $S(\widetilde{P})$ is given by

$$F \cdot \psi = \frac{1}{2} \sum_{i<j} R_{j,i} e_i e_j \cdot \psi + \frac{F_A}{2} \cdot \psi.$$

Fix an orthonormal basis $(e_1, \ldots, e_n)$ at a point. We write $\nabla_A = \nabla_{e_k} \otimes de_k$ at this point. We extend the e_k to commuting vector fields so that

$\nabla_{e_k}(e_\ell) = 0$ for all k, ℓ. (For example, use gaussian normal coordinates centered at the point.) We compute:

$$\sum_k e_k \nabla_{e_k} \left(\sum_\ell e_\ell \nabla_{e_\ell}(\psi) \right) = \tag{5.2}$$

$$= \sum_{k,\ell} e_k e_\ell \nabla_{e_k} \circ \nabla_{e_\ell}(\psi)$$

$$= \sum_k -\nabla_{e_k} \nabla_{e_k}(\psi) + \sum_{k<\ell} e_k e_\ell \left(\nabla_{e_k} \circ \nabla_{e_\ell} - \nabla_{e_\ell} \circ \nabla_{e_k} \right)(\psi)$$

$$= \sum_k -\nabla_{e_k} \nabla_{e_k}(\psi) + \sum_{k<\ell} e_k e_\ell F^{k,\ell}(\psi)$$

$$= \sum_k -\nabla_{e_k} \nabla_{e_k}(\psi) + \sum_{k<\ell} e_k e_\ell \left(\frac{1}{2} \sum_{i<j} R_{j,i}^{k,\ell} e_i e_j + \frac{F_A^{k,\ell}}{2} \right) \cdot \psi$$

Let us consider the first term in Equation 5.2

Claim 5.1.6
$$\nabla_A^* \nabla_A(\psi) = -\sum_k \nabla_{e_k} \nabla_{e_k}(\psi).$$

Proof.   First let us show that that on one-forms with values in $S(\widetilde{P})$ we have $\nabla_A^* = - * \nabla_A *$ where $*$ is the Hodge star operator. Let α be a one-form with values in $S(\widetilde{P})$ and φ a section of $S(\widetilde{P})$. Then we have by definition

$$\langle \nabla_A \varphi, \alpha \rangle_{L^2} = \langle \varphi, \nabla_A^*(\alpha) \rangle_{L^2}$$

On the other hand, since the connection preserves the metric, we see that

$$d \left(\langle \varphi(x), *\alpha(x) \rangle \right) = \langle \nabla_A(\varphi(x), *\alpha(x)) \rangle + \langle \varphi(x), \nabla_A(*\alpha)(x) \rangle.$$

Hence,

$$\langle \nabla_A(\varphi), \alpha \rangle_{L^2} = \int_X \langle \nabla_A \varphi(x), *\alpha(x) \rangle$$

$$= -\int_X \langle \varphi(x), \nabla_A(*\alpha)(x) \rangle$$

$$= \langle \varphi, - * \nabla_A(*\alpha) \rangle_{L^2}.$$

We conclude that for all φ and α we have

$$\langle \varphi, \nabla_A^*(\alpha) \rangle_{L^2} = \langle \varphi, - * \nabla_A(*\alpha) \rangle_{L^2}.$$

It is then clear that $\nabla_A^* = - * \nabla_A *$ on one-forms with values in $S(\tilde{P})$.
Now we apply this to the claim. We have

$$\nabla_A^* \nabla_A(\psi) = - * \nabla_A \left(\sum_k \nabla_{e_k}(\psi) * (de_k) \right)$$

$$= - * \sum_k \nabla_{e_k} \nabla_{e_k}(\psi) dvol$$

$$= - \sum_k \nabla_{e_k} \nabla_{e_k}(\psi).$$

$\square$

Now let us consider the second term in Equation 5.2. By skew symmetry
of R in both the upper and lower indices, we can rewrite the second term
in the last line above as:

$$\sum_{k<\ell} e_k e_\ell \left(\frac{1}{2} \sum_{i<j} R_{j,i}^{k,\ell} e_i e_j \psi \right) = \frac{1}{8} \sum_{k,\ell,i,j} R_{j,i}^{k,\ell} e_k e_\ell e_i e_j \psi$$

$$= \frac{-1}{8} \sum_{k,\ell,i,j} R_{i,j}^{k,\ell} e_k e_\ell e_i e_j \psi$$

Now let us break this term into various pieces:

$$\sum_{k,\ell,i,j} R_{i,j}^{k,\ell} e_k e_\ell e_i e_j \psi =$$

$$= \sum_j \left(\widetilde{\sum_{k,\ell,i}} R_{i,j}^{k,\ell} e_k e_\ell e_i + \sum_{k,\ell} R_{k,j}^{k,\ell} e_k e_\ell e_k + R_{\ell,j}^{k,\ell} e_k e_\ell e_\ell \right) e_j \psi$$

where the first sum is over distinct k, ℓ, i. The first term vanishes by
Lemma 5.1.1. Each of the second and third term simplifies to

$$\sum_j \sum_{k,\ell} R_{k,j}^{k,\ell} e_\ell e_j.$$

Adding these terms together and using the fact that the Ricci curvature is
symmetric in ℓ, j, the result is

$$2 \sum_{k,j} R_{k,j}^{k,j} e_j e_j \psi = -2 \sum_{k,j} R_{k,j}^{k,j} \psi = -2\kappa\psi.$$

Thus, the second term in Equation 5.2 evaluates to $\frac{\kappa}{4}\psi$.

Lastly, let us consider

$$\sum_{k<\ell} \frac{F_A^{k,\ell}}{2} e_k e_\ell \psi.$$

This is the definition of the action of the two-form $F_A/2$ via Clifford multiplication on the spinor field ψ.

Plugging in these three evaluations we see that

$$\partial_A \partial_A(\psi) = \nabla_A^* \nabla_A(\psi) + \frac{\kappa}{4}\psi + \frac{F_A}{2} \cdot \psi \tag{5.3}$$

as claimed.   □

Corollary 5.1.7 *Let (A, ψ) be a solution to the Seiberg-Witten equations for a $Spin^c$ structure $\tilde{P}$ on a compact riemannian four-manifold X. Then*

$$\|\nabla_A(\psi)\|_{L^2}^2 + \frac{1}{4}\langle \kappa\psi, \psi\rangle_{L^2} + \frac{\|\psi\|_{L^4}^4}{4} = 0.$$

In particular, setting κ_X^- equal to the maximum over $x \in X$ of $(0, -\kappa(x))$ we see that

$$\kappa_X^- \|\psi\|_{L^2}^2 \geq \|\psi\|_{L^4}^4.$$

Proof. Since ψ is a harmonic spinor from Equation 5.1 we have

$$0 = \partial_A \partial_A(\psi) = \nabla_A^* \nabla_A(\psi) + \frac{\kappa}{4}\psi + \frac{F_A}{2} \cdot \psi.$$

Using the fact that ψ is a plus spinor we see that $F_A \cdot \psi = F_A^+ \cdot \psi$. Thus, using the first of the Seiberg-Witten equations, we can rewrite the above equation as

$$0 = \nabla_A^* \nabla_A(\psi) + \frac{\kappa}{4}\psi + \frac{1}{2}\left(\psi \otimes \psi^* - \frac{|\psi|^2}{2}\mathrm{Id}\right)\psi.$$

This simplifies to

$$0 = \nabla_A^* \nabla_A(\psi) + \frac{\kappa}{4}\psi + \frac{|\psi|^2}{4}\psi.$$

Now let us take L^2-inner product with ψ and integrate over X. The result is

$$\|\nabla_A(\psi)\|_{L^2}^2 + \frac{1}{4}\langle \kappa\psi, \psi\rangle_{L^2} + \frac{\|\psi\|_{L^4}^4}{4} = 0.$$

□

We now obtain the following corollary immediately from this result.

Corollary 5.1.8 *If X is a compact riemannian four-manifold with non-negative scalar curvature then all solutions to the Seiberg-Witten equations (for any $Spin^c$-structure on X) have trivial spinor field ψ.*

5.2 A Priori bounds

Now we can use Equation 5.1 to establish an a priori pointwise bound to the size of the spinor field of any solution to the Seiberg-Witten equations.

First notice that since the C^∞ pairs (A, ψ) are dense in the configuration space, every configuration (A', ψ') is gauge equivalent to a pair (A'', ψ'') which lies in the slice through a C^∞ solution. If we restrict the Seiberg-Witten equations to the slice through a C^∞ object, then they become an elliptic equation with C^∞ coefficients. Thus, every solution to these equations contained in a slice through a C^∞ pair is in fact C^∞. This proves:

Lemma 5.2.1 *Every solution to the Seiberg-Witten equations is gauge equivalent to a C^∞ solution.*

Corollary 5.2.2 *Let X be a compact riemannian four-manifold. Define $\kappa^- : X \to \mathbf{R}$ by $\kappa^-(x) = \max(-\kappa(x), 0)$, and let $\kappa_X^- = \max_{x \in X} \kappa^-(x)$. Let $\widetilde{P} \to X$ be a $Spin^c$ structure. Suppose that (A, ψ) is a solution to the Seiberg-Witten equations for this $Spin^c$ structure. Then for every $x \in X$ we have*

$$|\psi(x)|^2 \le \kappa_X^-.$$

Proof. Since any solution (A, ψ) is gauge equivalent to a C^∞ solution and since the inequality that we wish to establish is invariant under changes of gauge, it suffices to consider the case of C^∞ solutions, which we now do.

Let us consider a point $x_0 \in X$ at which $|\psi(x)|^2$ achieves its maximum. We shall show that

$$|\psi(x_0)|^2 \le \kappa^-(x_0).$$

Clearly, the lemma is established once we show this inequality.

We begin with the fact that (A, ψ) is a C^∞ solution to the Seiberg-Witten equations. According to Equation 5.1 this means that for any $x \in X$ we have

$$\nabla_A^* \nabla_A (\psi(x)) + \frac{\kappa(x)}{4} \psi(x) + \frac{|\psi(x)|^2}{4} \psi(x) = 0.$$

Taking pointwise inner product with ψ yields

$$\langle \nabla_A^* \nabla_A(\psi(x)), \psi(x)\rangle + \frac{\kappa(x)}{4}|\psi(x)|^2 + \frac{|\psi(x)|^4}{4} = 0.$$

In particular, $\langle \nabla_A^* \nabla_A(\psi(x)), \psi(x)\rangle$ is real. Now we also have

$$-\sum_i \frac{\partial^2}{\partial e_i^2}\langle \psi(x), \psi(x)\rangle = -\sum_i \langle \nabla_{e_i} \circ \nabla_{e_i}(\psi(x)), \psi(x)\rangle$$

$$-2\sum_i \langle \nabla_{e_i}(\psi(x)), \nabla_{e_i}(\psi(x))\rangle$$

$$-\sum_i \langle \psi(x), \nabla_{e_i} \circ \nabla_{e_i}(\psi(x))\rangle$$

This implies that

$$\Delta(|\psi(x)|^2) + 2|\nabla_{e_i}(\psi(x))|^2 =$$
$$= \langle \nabla_A^* \circ \nabla_A(\psi(x), \psi(x)\rangle + \langle \psi(x), \nabla_A^* \circ \nabla_A(\psi(x)\rangle$$
$$= 2\mathrm{Re}\left(\langle \nabla_A^* \circ \nabla_A(\psi(x), \psi(x)\rangle\right).$$

Here Δ represents the Laplacian on functions. If x_0 is a local maximum of $|\psi(x)|^2$, then $\Delta(|\psi(x_0)|^2) \geq 0$. Hence, assuming that x_0 is such a local max, we obtain

$$\mathrm{Re}\langle \nabla_A^* \circ \nabla_A(\psi(x_0)), \psi(x_0)\rangle \geq 0.$$

But we have already seen that $\langle \nabla_A^* \circ \nabla_A(\psi(x_0)), \psi(x_0)\rangle$ is real. Thus, it is ≥ 0.

Thus, at a point $x_0 \in X$ where $|\psi(x)|^2$ achieves its maximum we have

$$\frac{\kappa(x_0)}{4}|\psi(x_0)|^2 + \frac{|\psi(x_0)|^4}{4} \leq 0.$$

It then follows either $|\psi(x_0)|^2 = 0$, which would mean that ψ is identically zero, or

$$|\psi(x_0)|^2 \leq \kappa^-(x_0).$$

This establishes the lemma.   □

Corollary 5.2.3 *Let (A, ψ) be a solution to the Seiberg-Witten equations. Then for any point $x \in X$ we have*

$$|F_A^+(x)| \leq \kappa_X^-/2.$$

Proof. By the curvature equation we know that

$$F_A^+ = \psi \otimes \psi^* - \frac{|\psi|^2}{2}\mathrm{Id}$$

and hence

$$|F_A^+(x)| = |\psi(x)|^2/2.$$

The corollary is now immediate from the previous one.   □

These estimates lead us to our first 'compactness' result which gives a bound on the number of $Spin^c$-structures for which there are solutions to the Seiberg-Witten equations.

Theorem 5.2.4 *Given a riemannian metric on X there are only finitely many $Spin^c$ structures up to isomorphism for that metric such that the moduli space of solutions to the Seiberg-Witten equations for the $Spin^c$ structure is non-empty and has non-negative formal dimension. For any solution (A, ψ) to the Seiberg-Witten equations at which the formal dimension is non-negative and for any $x \in X$ we have*

$$
\begin{aligned}
|\psi(x)|^2 &\leq \kappa_X^- \\
\|\nabla_A(\psi)\|_{L^2}^2 &\leq \frac{(\kappa_X^-)^2}{4}\mathrm{vol}(X) \\
|F_A^+(x)| &\leq \kappa_X^-/2 \\
\|F_A^+\|_{L^2}^2 &\leq (\kappa_X^-/2)^2\mathrm{vol}(X) \\
\|F_A^-\|_{L^2}^2 &\leq (\kappa_X^-/2)^2\mathrm{vol}(X) - 8\pi^2\chi(X) - 12\pi^2\,\mathrm{sign}(X).
\end{aligned}
$$

Proof. The first and third inequalities were just established in the previous two corollaries. The second inequality follows from the first equation in Corollary 5.1.7 and the first. The fourth follows immediately from the third by integrating over X. The last inequality we derive from the fourth and the fact that, since we are assuming that the formal dimension is non-negative, we have

$$c_1(\mathcal{L})^2 - (2\chi(X) + 3\,\mathrm{sign}(X)) \geq 0.$$

Of course,

$$c_1(\mathcal{L})^2 = \frac{1}{4\pi^2}\left(\|F_A^+\|_{L^2}^2 - \|F_A^-\|_{L^2}^2\right)$$

and the last inequality is easily established.

Now let us show that there are only finitely many $Spin^c$ structures up to isomorphism, for which the moduli space is non-empty and the formal dimension is non-negative. If (A, ψ) is a solution of the Seiberg-Witten equations at which the formal dimension of the moduli space is non-negative, then we have a bound depending only on the riemannian structure on both $\|F_A^+\|_{L^2}^2$ and $\|F_A^-\|_{L^2}^2$. This means that the cohomology class represented by $iF_A/(2\pi)$ lies in a compact subset inside $H^2(X; \mathbf{R})$. Since this class must also be integral, that implies that there are only finitely many possibilities for $c_1(\mathcal{L})$, the first Chern class of the determinant line bundle of the $Spin^c$ structure. Of course, there are only finitely many $Spin^c$ structures up to isomorphism whose determinant line bundle has a given first Chern class. $\square$

5.3   The Compactness of the Moduli Space

Now we fix a $Spin^c$ structure $\widetilde{P} \to X$ and we show that the moduli space of all solutions to the Seiberg-Witten equations for $\widetilde{P}$ is a compact space.

The Gauge-Fixing Lemma   First, we need a gauge fixing lemma which is in the same vein as Uhlenbeck's lemma for $SU(2)$-bundles but is much simpler since the equations are linear in our context.

Lemma 5.3.1 *Let $\mathcal{L}$ be a complex line bundle over X with a hermitian metric. Fix a unitary C^∞ connection A_0 on $\mathcal{L}$. Then for any $\ell \geq 0$ there are constants $K, C > 0$ depending only on X, A_0, ℓ such that the following hold. For any connection L_ℓ^2 unitary connection A on $\mathcal{L}$ there is an $L_{\ell+1}^2$ change of gauge σ such that $\sigma^*(A) = A_0 + \alpha$ where $\alpha \in L_\ell^2(T^*X \otimes i\mathbf{R})$ satisfies:*

$$d^*\alpha = 0$$

and

$$\|\alpha\|_{L_\ell^2}^2 \leq C\|F_A^+\|_{L_{\ell-1}^2}^2 + K.$$

Proof.   We write $A = A_0 + \alpha_0$ for $\alpha_0 \in L_\ell^2(T^*X \otimes i\mathbf{R})$. The element $d^*(\alpha_0) \in L_{\ell-1}^2(i\mathbf{R})$ is L^2-orthogonal to the constant functions. On this orthogonal complement $\mathcal{I}$ there is a continuous linear operator

$$\Delta^{-1} \colon \mathcal{I}_{L_{\ell-1}^2} \to \mathcal{I}_{L_{\ell+1}^2}$$

which inverts the restriction of the Laplacian to $\mathcal{I}$. Let

$$s_0 = \frac{-1}{2}\Delta^{-1}(d^*(\alpha_0)) \in L^2_{\ell+1}(i\mathbf{R}),$$

and let $\sigma_0 = \exp(s_0)$. Clearly, σ_0 is an $L^2_{\ell+1}$ change of gauge. Let $\alpha_1 = \alpha_0 + 2d\sigma_0 \in \Omega^1(X; i\mathbf{R})$. We have

$$(\det \sigma_0)^* A = A_0 + \alpha_1.$$

It is also clear that

$$d^*(\alpha_1) = d^*(\alpha_0) - d^* d\Delta^{-1} d^*(\alpha_0) = 0.$$

At this point we have shown how to gauge fix so that A becomes $A_0+\alpha_1$ with $d^*(\alpha_1) = 0$. Next, let us take up the estimates. The linear operator

$$(d^*, d^+)\colon L^2_\ell (T^* X \otimes i\mathbf{R}) \to L^2_{\ell-1}\left(i\mathbf{R} \oplus (\Lambda^2_+ T^* X \otimes i\mathbf{R})\right)$$

is an elliptic linear operator whose kernel is the harmonic one-forms. Thus, we can decompose α_1 into (h, β) where h is a harmonic one-form and β is L^2-orthogonal to the harmonic one-forms. Furthermore, for any one-form b orthogonal to the harmonic one-forms, there is a constant C depending only on X and ℓ such that

$$\|b\|^2_{L^2_\ell} \leq C\|(d^*(b), d^+(b))\|^2_{L^2_{\ell-1}}.$$

In our case of course, $d^*(\beta) = 0$ and $d^+(\beta) = F^+_A - F^+_{A_0}$. This means that

$$\|\beta\|^2_{L^2_\ell} \leq C\|F^+_A - F^+_{A_0}\|^2_{L^2_{\ell-1}} \leq C\|F^+_A\|^2_{L^2_{\ell-1}} + C\|F^+_{A_0}\|^2_{L^2_{\ell-1}}.$$

Setting $K_1 = \|F^+_{A_0}\|^2_{L^2_{\ell-1}}$, we establish

$$\|\beta\|^2_{L^2_\ell} \leq C\|F^+_A\|^2_{L^2_{\ell-1}} + CK_1.$$

Now we need to consider the harmonic projection h of α_1. There is no need for this term to be bounded, but we have further gauge transformations that we can apply in order to make this component bounded, without affecting β.

Claim 5.3.2 *For any purely imaginary harmonic one-form h_0 whose periods all lie in $(2\pi/i)\mathbf{Z}$ there is a harmonic function $\varphi\colon X \to S^1$ with $d\varphi = h_0$.*

Proof.   Choose a base point $x_0 \in X$ and integrate h_0 along paths to define a C^∞-function $\widetilde{\varphi} \colon \widetilde{X} \to i\mathbf{R}$ on the universal covering of X. Since the periods of h_0 are contained in $(2\pi/i)\mathbf{Z}$ it follows that $\widetilde{\varphi}$ descends to a function

$$\varphi \colon X \to i\mathbf{R}/2\pi i\mathbf{Z} = S^1.$$

Clearly, $d\varphi = h_0$, and hence $d^*d\varphi = 0$, so that φ is a harmonic function.   □

Since the quotient of the space of purely imaginary harmonic one-forms by those with periods in $(4\pi/i)\mathbf{Z}$ is a compact torus, there is a constant K_2, depending on ℓ, such that any purely imaginary harmonic one-form h can be written as $h_1 + 2h_2$ where h_2 has periods in $(2\pi/i)\mathbf{Z}$ and $\|h_1\|_{L^2_\ell} \leq K_2$. Applying this to the harmonic projection h of α_1 we write

$$h = h_1 - 2d\varphi$$

where $\varphi \colon X \to S^1$ is a harmonic function and $\|h_1\|_{L^2_\ell} \leq K_2$. Clearly,

$$(\det \varphi)^*(A_0 + \alpha_1) = A_0 + \alpha_1 + 2d\varphi = A_0 + h_1 + \beta.$$

Setting $\alpha = h_1 + \beta$ we have

$$(\det \varphi)^*(\sigma_0)^* A = A_0 + \alpha$$

with $d^*(\alpha) = 0$ and

$$\|\alpha\|_{L^2_\ell}^2 \leq \|h_1\|_{L^2_\ell}^2 + \|\beta\|_{L^2_\ell}^2 \leq K_2 + CK_1 + C\|F_A^+\|_{L^2_{\ell-1}}^2.$$

Setting $K = K_2 + CK_1$, this completes the proof of the gauge fixing lemma.
□

Bounds on dF_A^+ and on A.   The next step in the proof of compactness is to establish *a priori* bounds on the L^2-norm of dF_A^+ for any solution (A, ψ) to the Seiberg-Witten equations.

Lemma 5.3.3 *There is a constant C depending only on X such that for any solution (A, ψ) to the Seiberg-Witten equations we have*

$$\|dF_A^+\|_{L^2}^2 \leq C.$$

Proof.   Let $\nabla_{L.C.}$ denote the Levi-Cività connection

$$\nabla_{L.C.} \colon \Omega^0(X; \Lambda_+^2 T^*X) \to \Omega^1(X; \Lambda_+^2 T^*X).$$

By the Seiberg-Witten equations we have

$$F_A^+ = \psi \otimes \psi^* - \frac{|\psi|^2}{2}\mathrm{Id}.$$

Then we have

$$\nabla_{L.C.}F_A^+ = \nabla_A(\psi) \otimes \psi^* + \psi \otimes \nabla_A(\psi^*) - \mathrm{Re}\langle\nabla_A(\psi),\psi\rangle\mathrm{Id}.$$

But we have already seen that there a priori bounds on $\|\nabla_A(\psi)\|_{L^2}^2$ and the L^∞ norm of ψ. Clearly, then this gives an a priori bound on $\|\nabla_{L.C.}F_A^+\|_{L^2}^2$. But it is an easy exercise using the fact that the Levi-Cività connection is torsion-free to show that the composition

$$\Omega_+^2(X;\mathbf{R}) = \Omega^0(X;\Lambda_+^2 T^*X) \xrightarrow{\nabla_{L.C.}} \Omega^1(X;\Lambda_+^2 T^*X) \to \Omega^3(X;\mathbf{R})$$

(where the last map is obtained by skew symmetrizing) is equal to exterior d. From this it is clear that a bound on $\|\nabla_{L.C.}F_A^+\|_{L^2}^2$ implies a bound on $\|dF_A^+\|_{L^2}^2$ □

Corollary 5.3.4 *There is a constant C_1 depending only on X such that for any solution (A,ψ) to the Seiberg-Witten equations we have*

$$\|F_A^+\|_{L_1^2}^2 \leq C_1.$$

Proof. By elementary Hodge theory, for each $\ell \geq 0$ there is a constant $C' > 0$ such that for any self-dual two form F we have

$$\|\Pi(F)\|_{L_\ell^2}^2 \leq C\|dF\|_{L_{\ell-1}^2}^2$$

where Π denotes the L^2-projection orthogonal to the space of harmonic self-dual two forms. Thus, if we decompose $F_A^+ = H^+ + B$ with H^+ being a self-dual harmonic form and with B being L^2-orthogonal to the space of harmonic forms. We have

$$\|B\|_{L_1^2}^2 \leq C'\|dF_A^+\|_{L^2}^2 \leq C'C.$$

On the other hand we have an a priori L^2-bound on F_A^+ which implies that there is a constant $C'' > 0$ such that the projection H^+ of F_A^+ into the harmonic forms has L_1^2-norm squared at most C''. Setting $C_1 = C'C + C''$ establishes the lemma. □

Putting this together with the gauge fixing lemma from the previous paragraph we have the following bound.

Corollary 5.3.5 *Let $\widetilde{P}$ be a $Spin^c$ structure and let A_0 be a fixed C^∞-connection on $\mathcal{L} = \det(\widetilde{P})$. There is a constant K_1 depending only on X and A_0 such that for any solution (A, ψ) to the Seiberg-Witten equations we have a connection $A' = A_0 + \alpha$ gauge equivalent to A with*

$$d^* \alpha = 0$$

and

$$\|\alpha\|_{L_{\frac{3}{2}}^2}^2 \leq K_1.$$

C^∞ bounds on A and ψ. Now we are ready to parlay the L_2^2-bound on A and the L^∞ bound on ψ into C^∞ bounds on both A and ψ. The technique is the standard bootstraping technique for elliptic equations.

Theorem 5.3.6 *Suppose that (A, ψ) is a solution to the Seiberg-Witten equation and that we have fixed gauge so that $A = A_0 + \alpha$, where A_0 is a fixed C^∞ connection on the determinant line bundle, with $d^* \alpha = 0$ and with the projection of α into the harmonic forms contained in a given compact fundamental domain modulo the lattice of harmonic forms with periods in $(4\pi/i)\mathbf{Z}$. For every $\ell \geq 2$ there is a constant $C(\ell)$, depending only on X, A_0 and ℓ such that*

$$\|\alpha\|_{L_\ell^2}^2 + \|\psi\|_{L_\ell^2}^2 \leq C(\ell)$$

where the L_ℓ^2-norm of the spinor field in taken with respect to ∇_{A_0}.

Proof. We know already that ψ is pointwise bounded and that α is bounded in L_2^2. We also have a bound on $\|\nabla_A \psi\|_{L^2}^2$. Since we have an L_2^2 bound on α, this produces an L_1^2 bound on ψ (recall that in defining the norms of ψ we are taking derivatives with respect to A_0 not A).

Next let us show that ψ is bounded in L_3^2. By the Dirac equation we have

$$\partial\!\!\!/_{A_0}(\psi) = -\alpha \cdot \psi.$$

Since α is bounded in L_2^2 and ψ is pointwise bounded, it follows that $\partial\!\!\!/_{A_0}(\psi)$ is bounded in L^4. Since $\partial\!\!\!/_{A_0}$ is an elliptic operator, this implies that the L^2-projection of ψ orthogonal to the kernel of $\partial\!\!\!/_{A_0}$ is bounded in L_1^4. Since ψ is bounded L^2, its L^2-projection onto the kernel of $\partial\!\!\!/_{A_0}$ is also bounded in L^2 and hence is bounded C^∞. Together, these yield an L_1^4 bound on ψ.

Again using the Sobolev multiplication $L_2^2 \otimes L_1^4 \to L_1^3$ we see that $\partial\!\!\!/_{A_0}(\psi)$ is bounded in L_1^3 and hence that ψ is bounded in L_2^3. Arguing in the same way, using the Sobolev multiplication $L_2^2 \otimes L_2^3 \to L_2^2$, we see that $\partial\!\!\!/_{A_0}(\psi)$ is bounded in L_2^2 and hence that ψ is bounded in L_3^2. Once we know that ψ is bounded in L_3^2, it follows from the curvature equation and the Sobolev multiplication $L_3^2 \otimes L_3^2 \to L_3^2$ that F_A^+ is bounded in L_3^2 and hence by the gauge fixing lemma that α is bounded in L_4^2.

Now we come to the homogeneous part of the bootstrapping argument. Suppose by induction that for some $\ell \geq 3$ we have bounds for the L_ℓ^2-norms of α and ψ. Then from the Dirac equation and the continuous multiplication $L_\ell^2 \otimes L_\ell^2 \to L_\ell^2$ we see that $\partial\!\!\!/_{A_0}(\psi)$ is bounded in L_ℓ^2 and hence that ψ is bounded in $L_{\ell+1}^2$. From the curvature equation we conclude that F_A^+ is bounded in L_ℓ^2 and hence by the gauge fixing lemma that α is bounded in $L_{\ell+1}^2$. This completes the induction. $\qquad\square$

Now using the compactness of the embeddings $L_\ell^2 \subset C^{\ell-3}$ we have the following strong compactness result.

Corollary 5.3.7 *Let (A_n, ψ_n) be any sequence of solutions to the Seiberg-Witten equations. Then after passing to a subsequence, and applying L_3^2 changes of gauge we can arrange that the (A_n, ψ_n) are C^∞ objects and they converge in the C^∞ topology to a limit (A, ψ) which is also a solution to the Seiberg-Witten equations. In particular, the moduli space of solutions to the Seiberg-Witten equations is compact.*

We also have established the independence of the moduli space of the norm that we choose.

Corollary 5.3.8 *For any $\ell \geq 2$ let $\mathcal{C}_\ell(\widetilde{P})$ be the configuration space of L_ℓ^2 pairs (A, ψ), and let $\mathcal{G}_{\ell+1}(\widetilde{P})$ be the group of $L_{\ell+1}^2$ changes of gauge. We form the quotient space $\mathcal{B}_\ell(\widetilde{P})$ and the open subset $\mathcal{B}_\ell^*(\widetilde{P})$ of equivalence classes of irreducible pairs. Then $\mathcal{B}_\ell^*(\widetilde{P})$ is a Hilbert manifold. Let $\mathcal{M}_\ell(\widetilde{P}) \subset \mathcal{B}(\widetilde{P})$ be the moduli space of equivalence classes of solutions to the Seiberg-Witten equations. Then the natural map*

$$\iota_\ell \colon \mathcal{B}_\ell(\widetilde{P}) \to \mathcal{B}(\widetilde{P})$$

is an inclusion and is a smooth embedding on the open subset of irreducible classes. Furthermore, ι_ℓ induces a homeomorphism from $\mathcal{M}_\ell(\widetilde{P})$ to $\mathcal{M}(\widetilde{P})$. At any irreducible solution $[A, \psi] \in \mathcal{M}_\ell(\widetilde{P})$ the differential of ι_ℓ induces an isomorphism between the Zariski tangent spaces of the moduli spaces. In particular, the open subset of irreducible, smooth points of $\mathcal{M}_\ell(\widetilde{P})$ maps diffeomorphically onto the open subset of irreducible, smooth points of $\mathcal{M}(\widetilde{P})$.

Chapter 6

The Seiberg-Witten Invariant

6.1 The Statement

In general, there is no reason to expect that the space of solutions to the Seiberg-Witten equations form a smooth manifold. The best that one can hope for is that generically they do. In the $SU(2)$-ASD connection context one uses the metric as a parameter and shows that for a generic metric the ASD moduli space is smooth, see Section 3.2 of Chapter III of [2]. In this context there is no known generic metrics theorem. Instead we use a different family of perturbations. We perturb the curvature equation by adding a purely imaginary self-dual two-form.

Theorem 6.1.1 *Suppose that $b_2^+(X) > 0$. Fix a metric on X. Then for a generic C^∞ self-dual real two-form h on X the following holds. For any $Spin^c$-structure $\widetilde{P}$ on X the moduli space $\mathcal{M}(\widetilde{P}, h) \subset \mathcal{B}(\widetilde{P})$ of gauge equivalence classes of pairs $[A, \psi]$ which are solutions to the perturbed Seiberg-Witten equations:*

$$(SW_h): \qquad F_A^+ = q(\psi) + ih$$

$$\partial\!\!\!/_A(\psi) = 0$$

form a smooth compact submanifold of $\mathcal{B}^*(\widetilde{P})$ of dimension

$$\frac{c_1(\mathcal{L})^2 - (2\chi(X) + 3\sigma(X))}{4}$$

where $\chi(X)$ and $\sigma(X)$ are, respectively, the Euler characteristic and the signature of X. In particular, if this dimension is negative for a $Spin^c$ structure $\widetilde{P}$, then there are no solutions to the perturbed Seiberg-Witten equations for $\widetilde{P}$.

There are several steps in the proof of this theorem. First, let us introduce the parametrized moduli space.

6.2   The Parametrized Moduli Space

Fix a $Spin^c$-structure $\widetilde{P}$ for X. We shall study the parametrized moduli space for $\widetilde{P}$, parametrized by all self-dual two-forms and prove that this space is smooth. It is not convenient to work with C^∞-forms but rather with a Sobolev space of forms. We need the pointwise bounds of Chapter 5 to hold for the solutions to the perturbed equations, so this requires that the perturbations be continuous. We choose to allow the perturbing form to be L^2_3 (though we could work with a stronger norm). In order to have an elliptic theory we need to study L^2_4-configurations and L^2_5-changes of gauge. Let $\mathcal{C}_4(\widetilde{P})$ be the space of L^2_4-configurations and let $\mathcal{G}_5(\widetilde{P})$ be the group of L^2_5-changes of gauge.

We define a map

$$\mathcal{C}_4(\widetilde{P}) \times L^2_3\left(\Lambda^2_+ T^* X \otimes \mathbf{R}\right) \to L^2_3\left(\left(\Lambda^2_+ T^* X \otimes \mathbf{R}\right) \oplus S^-(\widetilde{P})\right)$$

by associating to (A, ψ, h) the element

$$F(A, \psi, h) = (F^+_A - q(\psi) - ih, \partial\!\!\!/_A(\psi)).$$

Lemma 6.2.1 *At any point (A, ψ, h) for which $F(A, \psi, h) = 0$ and $\psi \neq 0$ the differential DF of this map is onto.*

Proof.   To establish this claim first notice that the differential DF restricted to the h-coordinate is minus the inclusion of the first factor in the range. Hence, the restriction of DF to the second factor of the domain is onto the first factor in the range. We complete the proof of this claim by showing that the restriction of DF to the tangent space

$$L^2_4\left((T^* X \otimes i\mathbf{R}) \oplus S^+(\widetilde{P})\right)$$

to $\mathcal{C}_4(\widetilde{P})$ followed by projection onto the second factor of the range is onto. This composition is given by the map

$$G\colon (a, \eta) \mapsto \partial\!\!\!/_A(\eta) + \frac{1}{2}a \cdot \psi$$

where the multiplication in the second term is Clifford multiplication. We shall show that G is surjective. Suppose that an element $\lambda \in L_3^2\left(S^-(\widetilde{P})\right)$ which is L^2-orthogonal to the image of G. In particular, λ is orthogonal to the image of the Dirac operator $\partial\!\!\!/_A$ and hence, since $\partial\!\!\!/_A$ is self-adjoint, is in the kernel of the dirac operator on $S^-(\widetilde{P})$. We suppose that $\lambda \neq 0$. By elliptic regularity this means that λ does not vanish on any open subset of X. Similarly, since ψ is in the kernel of $\partial\!\!\!/_A$ and by hypothesis is non-zero, it also does not vanish on any open subset. Of course, both λ and ψ are continuous. We choose a small open ball $U \subset X$ centered at a point x_0 where both ψ and λ are non-zero. We make U sufficiently small so that both ψ and λ are almost constant over this ball with respect to a coordinate system on the ball. The projection mapping from $\mathbf{R}^4 \to Cl_1(\mathbf{R}^4)^+$ is an isomorphism. Of course, Clifford multiplication induces an isomorphism

$$\mathbf{R}^4 \otimes_{\mathbf{R}} \mathbf{C} \cong \left(Cl_1(\mathbf{R}^4) \otimes_{\mathbf{R}} \mathbf{C}\right)^- \xrightarrow{\cong} \mathrm{Hom}_{\mathbf{C}}(S^+(\mathbf{R}^4), S^-(\mathbf{R}^4)).$$

This means that given non-zero vectors $\sigma^\pm \in S^\pm(\mathbf{R}^4)$ there is an element $a \in \mathbf{R}^4$ with

$$\mathrm{Re}\left(\langle a \cdot \sigma^+, \sigma^- \rangle\right) > 0.$$

It follows that there is a vector in $a \in T^*X_{x_0}$ such that

$$\mathrm{Re}\left(\langle a \cdot \psi(x_0), \lambda(x_0) \rangle\right) > 0.$$

Extending a over all of U by using the trivialization of the cotangent bundle induced by the coordinate system, we find a one-form a on U such that at each point $x \in U$ we have

$$\mathrm{Re}\left(\langle a(x) \cdot \psi(x), \lambda(x) \rangle\right) > 0.$$

Damping a out to zero near the boundary of U we find a global one-form a on X, vanishing outside of U, such that

$$\int_X \mathrm{Re}\left(\langle a(x) \cdot \psi(x), \lambda(x) \rangle\right) dvol > 0.$$

This means that λ is not orthogonal to $G(2a, 0)$. This is a contradiction and establishes that the orthogonal complement to the image of

$$DF_{(A,\psi)} \colon L_4^2\left((T^*X \otimes i\mathbf{R}) \oplus S^+(\widetilde{P})\right) \to L_3^2\left((\Lambda_+^2 T^*X \otimes i\mathbf{R}) \oplus S^-(\widetilde{P})\right)$$

is trivial, and hence that this linear map is onto. This completes the proof of the claim. $\qquad\square$

Proposition 6.2.2 *Let X be a closed oriented riemannian four-manifold. Fix a $Spin^c$ structure $\tilde{P}$ for X. The parametrized moduli space of irreducible solutions; i.,e., the subspace*

$$\mathcal{PM}^*(\tilde{P}) \subset \mathcal{B}_4^*(\tilde{P}) \times L_3^2 \left(\Lambda_+^2 T^* X \otimes \mathbf{R} \right)$$

consisting of all $([A, \psi], h)$ for which $F_A^+ = q(\psi) + ih$ and $\partial_A(\psi) = 0$ forms a smooth submanifold. The projection of this submanifold to the parameter space $L_3^2 \left(\Lambda_+^2 T^* X \otimes \mathbf{R} \right)$ is a smooth mapping. The fiber over h is denoted $\mathcal{M}^*(\tilde{P}, h)$. It is the moduli space of irreducible solutions, modulo change of gauge, to the perturbed Seiberg-Witten equations SW_h. The kernel of the differential of the projection mapping at $([A, \psi], h)$ is the Zariski tangent space to the moduli space $\mathcal{M}^*(\tilde{P}, h)$ at $[A, \psi]$. By the transversality condition, the cokernel is identified with the obstruction space for $\mathcal{M}^*(\tilde{P}, h)$ at $[A, \psi]$. Thus, the differential of the projection mapping is Fredholm and its index is

$$d(\mathcal{L} = \frac{c_1(\mathcal{L})^2 - (2\chi(X) + 3\sigma(X))}{4}.$$

Proof. It follows from the claim that the subspace

$$\widetilde{\mathcal{PM}}^*(\tilde{P}) \subset \mathcal{C}_4^*(\tilde{P}) \times L_3^2 \left(\Lambda_+^2 T^* X \otimes \mathbf{R} \right)$$

consisting of all triples (A, ψ, h) for which $F_A^+ = q\psi + ih$ and $\partial_A(\psi) = 0$ forms a smooth submanifold. Dividing out by the action of the group of gauge transformations yields a smooth submanifold $\mathcal{PM}^*(\tilde{P})$ of $\mathcal{B}^*(\tilde{P}) \times L_3^2 \left(\Lambda_+^2 T^* X \otimes \mathbf{R} \right)$. The statements about the differential of the projection mapping are easily established. □

Corollary 6.2.3 *For a generic $h \in L_3^2 \left(\Lambda_+^2 T^* X \otimes \mathbf{R} \right)$ the space $\mathcal{M}^*(\tilde{P}, h)$ of irreducible solutions to the perturbed Seiberg-Witten equations SW_h is a smooth submanifold of $\mathcal{B}_4^*(\tilde{P})$ of dimension $d(\mathcal{L})$.*

Proof. By Smale-Sard theorem we conclude from the previous result that for a generic $h \in L_3^2 \left(\Lambda_+^2 T^* X \otimes \mathbf{R} \right)$ at every irreducible solution for the perturbed Seiberg-Witten equations SW_h, the linearization of the equations is surjective. Furthermore, the moduli space of all irreducible solutions of these equations is a smooth (infinite dimensional) manifold. Dividing out by the action of the group of gauge transformations (which acts freely since we are working at irreducible solutions), we see that the moduli space of gauge equivalence classes of solutions of the perturbed Seiberg-Witten equations SW_h is a smooth manifold whose index is given by minus the Euler characteristic of the fundamental elliptic complex $\mathcal{E}(A, \psi)$. $\quad\square$

6.3 Reducible Solutions

Now let us consider the issue of reducible solutions. This is where we use the condition that $b_2^+(X) > 0$.

Proposition 6.3.1 *Suppose that $b_2^+(X) > 0$. Let $\widetilde{P} \to X$ be a Spin^c-structure for which $c_1(\mathcal{L})$ is not a torsion cohomology class. Then for a generic metric there are no reducible solutions to the Seiberg-Witten equations. For any Spin^c structure, if there are no reducible solutions to the Seiberg-Witten equations then for all sufficiently small perturbations of the equations as in the last section there are no reducible solutions, and for any metric and a generic perturbation there are no reducible solutions.*

Proof. A reducible solution to the Seiberg-Witten equations is the same thing as an anti-self-dual connection on the determinant line bundle and a zero spinor field. Such an ASD connection has curvature which is a harmonic two-form whose cohomology class is $2\pi/i$ times the first Chern class of the line bundle. For this form to be ASD it must be orthogonal to the self-dual harmonic two-forms. According to the argument in [11], if $c_1(\mathcal{L}) \in H^2(X; \mathbf{R})$ is non-zero, then this condition is a closed codimension $b_2^+(X)$ condition in the space of all metrics.

For any metric, a reducible solution to the perturbed equations satisfies $F_A^+ = ih$. If the orthogonal projection of h into the self-dual harmonic two-forms is not equal to the projection of 2π times c_1 into the self-dual harmonic two-forms, then there is no such solution. From this the last two statements are immediate. $\quad\square$

Corollary 6.3.2 *Let X be a closed, smooth, oriented riemannian manifold with $b_2^+(X) > 0$. Then for a generic L_3^2-self dual real two-form h, the moduli space $\mathcal{M}(\widetilde{P}, h)$ is a smooth submanifold of $\mathcal{B}^*(\widetilde{P})$ of dimension $d(\mathcal{L})$.*

Proof. We have established that there are no reducible solutions to the perturbed equations and that at each irreducible solution the moduli space is smooth of the given dimension. $\quad\square$

6.4   Compactness of the Perturbed Moduli Space

Proposition 6.4.1 *Fix a closed, oriented, riemannian four-manifold X and a Spin^c structure $\widetilde{P}$ for X. For any $h \in L_3^2\left(\Lambda_+^2 T^* X \otimes \mathbf{R}\right)$ the moduli space $\mathcal{M}(\widetilde{P}, h)$ is compact.*

Proof.   In the case of a solutions of the perturbed equations the Bochner formula (Equation 5.3) tells us

$$0 = \nabla_A^* \nabla_A(\psi) + \frac{\kappa}{4}\psi + \frac{|\psi|^2}{4}\psi + ih \cdot \psi. \qquad (6.1)$$

Taking pointwise inner product with ψ, evaluating at a maximum x_0 for $|\psi|$, and arguing as in the proof of Corollary 5.2.2, we find

$$0 \geq \frac{\kappa(x_0)}{4}|\psi(x_0)|^2 + \frac{|\psi(x_0)|^4}{4} + \mathrm{Re}\left(\langle ih(x_0) \cdot \psi(x_0), \psi(x_0)\rangle\right).$$

It is clear that the last term in this expression has norm at most

$$|h(x_0)||\psi(x_0)|^2.$$

Thus, we conclude that either $\psi(x_0) = 0$, in which case ψ is identically zero, or

$$0 \geq \frac{\kappa}{4} + \frac{|\psi(x_0)|^2}{4} - |h(x_0)|.$$

It follows that if ψ is not identically zero, then for all $x \in X$ we have

$$|\psi(x)|^2 \leq 4|h(x_0)| - \kappa(x_0).$$

This means that for any solution to the perturbed equations we have

$$|\psi(x)|^2 \leq \max\left(\max_{y \in X}(4|h(y)| - \kappa(y)), 0\right).$$

This pointwise bound on ψ takes the place of the bound given in Corollary 5.2.2.

We also have an *a priori* bound on $\|\nabla_A(\psi)\|_{L^2}^2$. We obtain this bound by taking the L^2 inner product of Equation 6.1 with ψ. We obtain

$$\|\nabla_A(\psi)\|_{L^2}^2 + \frac{\kappa}{4}\|\psi\|_{L^2}^2 + \|\psi\|_{L^4}^4 + \langle ih\psi, \psi\rangle_{L^2} = 0.$$

Since κ, ψ, h are all pointwise bounded *a priori*, it follows that $\|\nabla_A(\psi)\|_{L^2}^2$ is bounded *a priori* as well.

Once we have these two *a priori* bounds the argument proceeds exactly as in Section 5.3 to show that the solutions are bounded in L_ℓ^2 for every ℓ, and hence that the moduli space to the perturbed equations is compact. $\square$

Corollary 6.4.2 *Now suppose that $b_2^+(X) > 0$ and fix a $Spin^c$ structure $\widetilde{P}$ for X. Then there is an open dense subset $U(\widetilde{P}) \subset L_3^2\left(\Lambda_+^2 T^* X \otimes \mathbf{R}\right)$ such that for any $h \in U(\widetilde{P})$ the moduli space $\mathcal{M}(\widetilde{P}, h)$ consists of irreducible solutions and is a compact smooth submanifold of $\mathcal{B}^*(\widetilde{P})$ of dimension $d(\mathcal{L})$.*

Proof. We have now shown that the parametrized moduli space of irreducible solutions

$$\mathcal{PM}^*(\widetilde{P}) \subset \mathcal{B}_4^*(\widetilde{P}) \times L_3^2\left(\Lambda_+^2 T^* X \otimes \mathbf{R}\right)$$

is a smooth submanifold and the projection of this smooth submanifold to the parameter space $L_3^2\left(\Lambda_+^2 T^* X \otimes \mathbf{R}\right)$ is a smooth mapping. Since $b_2^+(X) > 0$ there is a closed, nowhere dense subset of the parameter space for which there are reducible solutions to the perturbed Seiberg-Witten equations. Removing this subset from the parameter space and its preimage from the parametrized moduli space, the projection mapping to the parameter space has Fredholm differential and compact fibers. This means that there is an open dense subset of the parameter space consisting of regular values for the map. Since the moduli space of L_4^2-solutions to the perturbed equations modulo L_5^2-changes of gauge is equal to the moduli space of L_2^2-solutions modulo L_3^2-changes of gauge, this produces an open dense subset $U(\widetilde{P})$ of $L_3^2\left(\Lambda_+^2 T^* X \otimes \mathbf{R}\right)$ as required. $\square$

Proof of Theorem 6.1.1. For each $Spin^c$ structure $\widetilde{P}$. Let $U^\infty(\widetilde{P})$ be the intersection of the open subset $U(\widetilde{P}) \subset L_3^2\left(\Lambda_+^2 T^* X \otimes \mathbf{R}\right)$ of Corollary 6.4.2 with the subspace of C^∞-forms. Clearly, by Corollary 6.4.2, $U^\infty(\widetilde{P})$ is an open dense subset of the space of C^∞ self-dual two-forms on X and the conclusion of Theorem 6.1.1 holds for the $Spin^c$ structure $\widetilde{P}$ and each $h \in U^\infty(\widetilde{P})$. Now let U^∞ be the intersection over all $Spin^c$ structures $\widetilde{P}$ for X of $U^\infty(\widetilde{P})$. Since there are only countably many $Spin^c$ structures up to isomorphism and since $U^\infty(\widetilde{P})$ clearly depends only on the isomorphism class of the $Spin^c$ structure $\widetilde{P}$, it follows that U^∞ is a dense G_δ in the space of all C^∞ self-dual two-forms. It is also clear that the conclusion of Theorem 6.1.1 holds for any $h \in U^\infty$.

6.5   Variation of the Metric and Self-dual Two-form

We also need a corresponding result for generic paths of metrics and perturbations.

Theorem 6.5.1 *Suppose that X is a closed, smooth oriented 4-manifold and that $b_2^+(X) > 1$. Suppose that g_0, g_1 are metrics on X and that h_0, h_1 are generic self-dual two-forms on X such that Theorem 6.1.1 holds for both g_0, h_0 and g_1, h_1. Let $\gamma = \gamma(t)$ be a smooth path of metrics connecting g_0 to g_1 and let $\eta = \eta(t)$ be a generic path of C^∞-self-dual L_3^2 two-forms connecting h_0 and h_1. We define the parametrized moduli space $\mathcal{M}(\widetilde{P}, \eta)$ consisting of all*

$$([A, \psi], t) \subset \mathcal{B}(\widetilde{P}) \times [0, 1]$$

satisfying the equations

$$F_A^{+_t} = q(\psi) + i\eta(t)$$
$$\partial\!\!\!/_{A, g_t}(\psi) = 0$$

where $+_t$ means the self-dual projection with respect to g_t and where $\partial\!\!\!/_{A, g_t}$ means the Dirac operator constructed using the Levi-Cività connection associated to g_t and the connection A on $\mathcal{L}$. Then $\mathcal{M}(\widetilde{P}, \eta)$ consists only of irreducible points and is a smooth compact submanifold with boundary of $\mathcal{B}^*(\widetilde{P}) \times [0, 1]$ whose boundary is the disjoint union of the moduli spaces associated to g_0, h_0 and g_1, h_1.

Proof.   Let us define $\mathcal{P}(h_0, h_1)$ to be the space of L_1^2-maps η from $[0, 1]$ to $L_3^2\left(\Lambda_+^2 T^* X \otimes \mathbf{R}\right)$ satisfying $\eta(0) = h_0$ and $\eta(1) = h_1$. All such paths are continuous. This space is a Hilbert manifold whose tangent space at any point η is the space of L_1^2-functions from $[0, 1]$ to $L_3^2\left(\Lambda_+^2 T^* X \otimes \mathbf{R}\right)$ vanishing at the endpoints. We define a map

$$\mathbf{F} \colon \mathcal{C}_4^*(P) \times I \times \mathcal{P} \to L_3^2\left(\Lambda_+^2 T^* X \otimes i\mathbf{R} \oplus S^-(\widetilde{P})\right)$$

by

$$\mathbf{F}(A, \psi, t, \eta) = (F_A^{+_t} - q(\psi) - i\eta(t), \partial\!\!\!/_{A, g_t}(\psi), t).$$

We claim that the differential $D\mathbf{F}$ is surjective at every (A, ψ, t, η) for which $\mathbf{F}$ vanishes. Since h_0, h_1 are chosen generically, this is automatic for all solutions over $t = 0$ and $t = 1$. For solutions whose t coordinate satisfies

$0 < t < 1$ we know, by the same argument as in the absolute case that the differential from the tangent space of $\mathcal{P}$ to $L^2_3\left(\Lambda^2_+ T^* X \otimes i\mathbf{R}\right)$ is surjective. Once we have this, the same argument as in the absolute case then establishes the surjectivity. Let $\mathbf{M}^*(\widetilde{P})$ be the moduli space of irreducible solutions parametrized by $\mathcal{P}$. That is to say $\mathbf{M}^*(\widetilde{P})$ is the quotient of $\mathbf{F}^{-1}(0)$ by the action of $\mathcal{G}_5(\widetilde{P})$. By transversality $\mathbf{M}^*(\widetilde{P})$ is a smooth manifold with boundary. The projection of $\mathbf{M}^*(\widetilde{P})$ to $\mathcal{P}$ is a smooth map and hence by Sard's theorem for a generic $\eta \in \mathcal{P}$ the fiber over η is a smooth manifold with boundary. This fiber is exactly exactly the subset of irreducible solutions in $\mathcal{M}(\widetilde{P}, \eta)$.

For a generic path η of self-dual two-forms there will be no reducible solutions in $\mathcal{M}(\widetilde{P}, \eta)$ since the subset of perturbations h for which there are reducible solutions to the perturbed equations is a smooth submanifold of codimension at least two in the space of all self-dual two-forms. Thus, for a generic η the fiber of the projection from $\mathcal{M}^*(\widetilde{P}) \to \mathcal{P}$ is $\mathcal{M}(\widetilde{P}, \eta)$. The compactness of $\mathcal{M}(\widetilde{P}, \eta)$ follows directly from the compactness of each of the $\mathcal{M}(\widetilde{P}, \eta(t))$ which is a consequence of Proposition 6.4.1. At this point we have established that for a generic L^2_1-path η the parametrized moduli space $\mathcal{M}(\widetilde{P}, \eta)$ is a compact, properly embedded, smooth submanifold of $\mathcal{B}^*(\widetilde{P}) \times I$. Its boundary is the disjoint union of $\mathcal{M}(\widetilde{P}, h_1)$ and $\mathcal{M}(\widetilde{P}, h_0)$. Since the moduli spaces are compact, the η for which these statements hold is an open subset of the space of L^2_1-paths and hence is an open dense subset. Thus, there is a dense G_δ in the space of C^∞ paths from h_0 to h_1 for which the conclusion of the theorem holds. $\qquad\square$

6.6 Orientability of the Moduli Space

There is just one issue left to deal with before we can define the Seiberg-Witten invariant of a $Spin^c$ structure. That is the orientability of the moduli space.

If $F\colon V \to W$ is a Fredholm (linear) map, then we define the determinant line $\det(F)$ to be the top exterior power of the kernel of F tensored with the inverse of the top exterior power of the cokernel of F. If $\{F_s\}_{s \in S}$ is a continuous family of Fredholm operators, then there is a natural way to fit these determinant lines together to form a line bundle over the parameter space S, see for example Section 5.3 of Chapter IV of [2].

First an elementary lemma.

Lemma 6.6.1 *Let F_s be a continuous family of Fredholm operators para-*

metrized by $s \in S$. Suppose that F'_s is a homotopic family. Then the homotopy between the families determines an isomorphism between the determinant line bundles $L \to S$ and $L' \to S$ of these families. The isomorphism is well-defined up to the action of a positive real function on S.

Proof. We can view the homotopy between the families as a family of Fredholm operators over $S \times I$. We then have its determinant line bundle, which is a line bundle over $S \times I$. Any line bundle over $S \times I$ is isomorphic to the product with I of a line bundle over S. In particular, the restrictions of the line bundle to the two ends $S \times \{0\}$ and $S \times \{1\}$ are isomorphic by an isomorphism well-defined up to multiplication by a positive function. $\quad\square$

Now let us apply this to the family of Fredholm operators parametrized by the points of $\mathcal{C}(\tilde{P})$. To each $(A, \psi) \in \mathcal{C}(\tilde{P})$ we have the 'elliptic complex'

$$0 \xrightarrow{} L^2_3(X; i\mathbf{R}) \xrightarrow{D_0} L^2_2\left((T^*X \otimes i\mathbf{R}) \oplus S^+(\tilde{P})\right)$$
$$\xrightarrow{D_2} L^2_1\left((\Lambda^2_+ T^*X \otimes i\mathbf{R}) \oplus S^-(\tilde{P})\right) \xrightarrow{} 0$$

where the first map D_1 in the complex is $(2d, -(\cdot)\psi)$ and the second map D_2 is given by the matrix

$$\begin{pmatrix} P_+ d & -Dq_\psi \\ \cdot \frac{1}{2}\psi & \partial\!\!\!/_A \end{pmatrix}.$$

In point of fact, if (A, ψ) is not a solution to the Seiberg-Witten equations, then this sequence of operators is not necessarily a complex in the sense that $D_2 \circ D_1$ is not equal to zero. Nevertheless, we can make a Fredholm operator out of the sequence by forming (D_2, D_1^*) mapping

$$L^2_2\left((T^*X \otimes i\mathbf{R}) \oplus S^+(\tilde{P})\right) \to L^2_1\left((\Lambda^2_+ T^*X \otimes i\mathbf{R}) \oplus S^-(\tilde{P}) \oplus i\mathbf{R}\right).$$

We denote the map (D_2, D_1^*) by $D_{(A,\psi)}$. This is a smooth family of Fredholm operators parametrized by the points of $\mathcal{C}(\tilde{P})$. It is convenient to deform this family by a homotopy

$$D_{(A,\psi)}(t) = (D_2(t), D_1(t)^*)$$

where $D_1(t) = 2d - (\cdot)(1-t)\psi$ and $D_2(t)$ is given by the matrix

$$\begin{pmatrix} P_+ d & -(1-t)Dq_\psi \\ \cdot \frac{1-t}{2}\psi & \partial\!\!\!/_A \end{pmatrix}.$$

Clearly, this is a homotopy between the family $D_{(A,\psi)}$ at $t = 0$ and the family $E_{(A,\psi)}$ at $t = 1$ where $E_{(A,\psi)}$ is $d^+ + \partial\!\!\!/_A + 2d^*$ mapping

$$L_2^2\left((T^*X \otimes i\mathbf{R}) \oplus S^+(\widetilde{P})\right) \to L_1^2\left((\Lambda_+^2 T^*X \otimes i\mathbf{R}) \oplus S^-(\widetilde{P})\right) \oplus L_1^2(X; i\mathbf{R}).$$

The determinant of this family of operators is a real line bundle over $\mathcal{C}(\widetilde{P}) \times I$. The action of $\mathcal{G}(\widetilde{P})$ on $\mathcal{C}(\widetilde{P})$ lifts to an obvious action on the total space of this determinant line bundle, an action which is linear on each fiber. Thus, restricting to $\mathcal{C}^*(\widetilde{P}) \times I$ we can form the quotient by this action giving us a determinant line bundle ξ over $\mathcal{B}^*(\widetilde{P}) \times I$.

Claim 6.6.2 *The real line bundle ξ over $\mathcal{B}^*(\widetilde{P}) \times I$ is isomorphic to a trivial real line bundle. An orientation for this line bundle is determined by choosing an orientation of*

$$\Lambda^{\mathrm{top}} H^1(X; i\mathbf{R}) \otimes \left(\Lambda^{\mathrm{top}} H^2(X; i\mathbf{R})\right)^{-1} \otimes H^0(X; i\mathbf{R})^{-1}.$$

Proof. To prove that the line bundle is trivial it suffices to show that its restriction to $\mathcal{B}^*(\widetilde{P}) \times \{1\}$ is trivial. Over $\mathcal{C}^*(\widetilde{P} \times \{1\})$ the family of Fredholm operators splits as the direct sum of a constant operator $d^+ + 2d^*$ and a varying family of complex linear operators $\partial\!\!\!/_A$. Thus, over this subspace the determinant line bundle is the tensor product of a constant line, the determinant of the constant family of operators, and the real determinant line bundle of the varying family of complex operators. The action of $\mathcal{G}(\widetilde{P})$ preserves this decomposition and preserves the complex structure of the second factor. Thus, the quotient by the action of $\mathcal{G}(\widetilde{P})$ has the same type of tensor product decomposition. The real determinant line bundle of any family of complex linear operators is trivial and is naturally oriented by the complex structure. Of course, the constant family determines a trivial determinant line bundle. This proves that ξ is isomorphic to a trivial line bundle.

Since $\mathcal{B}^*(\widetilde{P}) \times I$ is connected, to orient ξ it suffices to orient it any point. We consider a point $(x, 1) \in \mathcal{B}^*(\widetilde{P}) \times \{1\}$. Clearly, from the above description, to orient the determinant line over $(x, 1)$ we need only orient the determinant line of the constant family. The kernel of the constant family is $H^1(X; i\mathbf{R})$, and the cokernel is $H_+^2(X; i\mathbf{R}) \oplus H^0(X; i\mathbf{R})$. The claim is now proved. $\square$

Corollary 6.6.3 *The open subset of smooth, irreducible points of $\mathcal{M}(\widetilde{P})$ is an orientable manifold. A choice of orientations of $H^0(X; i\mathbf{R})$, $H^1(X; i\mathbf{R})$,* and $H^2_+(X; i\mathbf{R})$ *determine an orientation of* $\mathcal{M}(\widetilde{P})$ *at any smooth, irreducible point.*

Proof.   A choice of orientations of the three cohomology groups determines a orientation of the determinant line bundle ξ over $\mathcal{B}(\widetilde{P})$ formed from the determinant line bundle of the family of Fredholm operators $D_{(A,\psi)}$ over $\mathcal{C}^*(\widetilde{P})$. Suppose that $[A, \psi] \in \mathcal{M}(\widetilde{P})$ is a smooth, irreducible point. Then the kernel of $D_{(A,\psi)}$ is equal to the tangent space to $\mathcal{M}(\widetilde{P})$ and the cokernel is trivial. Thus, the determinant of $D_{(A,\psi)}$ is naturally identified with the top exterior power of the tangent bundle to $\mathcal{M}(\widetilde{P})$ at $[(A, \psi]$. Since the determinant line bundle is orientable, it follows that the manifold of smooth, irreducible points of $\mathcal{M}(\widetilde{P})$ forms an orientable manifold. The rest of the corollary is immediate from the previous result. $\square$

In fact this argument applies to the perturbed equations and to any parametrized moduli space parametrized by an oriented manifold.

Corollary 6.6.4 *Let S be a smooth, finite dimensional, orientable manifold. For each $s \in S$ let g_s be a riemannian metric on on X and let h_s be a self-dual two-form. Suppose that both g_s and h_s vary smoothly with s. Let $\mathcal{M}(\widetilde{P}, S)$ be the parametrized moduli space of all $([A, \psi], s)$ such that (A, ψ) is a solution to the Seiberg-Witten equations for the metric g_s perturbed by the form ih_s. Suppose that this moduli space is a smooth manifold. Then a choice of orientations of $H^0(X; i\mathbf{R})$ of $H^1(X; i\mathbf{R})$ and of $H^2_+(X; i\mathbf{R})$ and for S determines an orientation for $\mathcal{M}(\widetilde{P}, S)$.*

Orientation Convention.   We explicitly give the convention for orienting the determinant line bundle from the orientations of $H^1(X; \mathbf{R})$ and $H^2_+(X; \mathbf{R})$. First, $H^0(X; \mathbf{R})$ is given its usual orientation: the cohomology class 1 determines the orientation. The orientation of the kernel of the constant operator $d^+ + 2d^*$ is simply given by the image of the chosen orientation of $H^1(X; \mathbf{R})$ under multiplication by i. (The kernel of this operator is $H^1(X; i\mathbf{R})$.) The orientation of the cokernel of this operator is the direct sum of the image under multiplication by i of the chosen orientation for $H^2_+(X; \mathbf{R})$ followed by the image under multiplication by i of the natural orientation for $H^0(X; \mathbf{R})$.

6.7 The Case when $b_2^+(X) > 1$

Let us fix a riemannian manifold X. We suppose that $b_2^+(X) > 1$. Fix a $Spin^c$ structure $\widetilde{P} \to X$ for X. Choose orientations for $H^1(X; \mathbf{R})$ and $H_+^2(X; \mathbf{R})$. Then for a generic C^∞ self-dual two-form h, the moduli space $\mathcal{M}(\widetilde{P}, h)$ of solutions to the perturbed Seiberg-Witten equations is a compact smooth submanifold of $\mathcal{B}^*(\widetilde{P})$. The orientations of the cohomology spaces determines an orientation of the moduli space.

We have seen (Section 4.5) that over $\mathcal{B}^*(\widetilde{P})$ there is a principal S^1-bundle whose total space consists of equivalence classes of irreducible configurations under the action of the group $\mathcal{G}^0(\widetilde{P})$ of based changes of gauge. Let $\mu \in H^2(\mathcal{B}^*(\widetilde{P}); \mathbf{Z})$ be the first Chern class of this principal S^1-bundle. If the dimension of this moduli space, which is

$$d(\mathcal{L}) = \left(c_1(\mathcal{L})^2 - 2\chi(X) - 3\sigma(X)\right)/4,$$

is even, say equal to $2d$, then we define the Seiberg-Witten invariant of $\widetilde{P}$ as follows:

$$SW(\widetilde{P}) = \int_{\mathcal{M}(\widetilde{P}, h)} \mu^d.$$

If the dimension of the moduli space is odd, then we define the Seiberg-Witten invariant of $\widetilde{P}$ to be zero.

Note. The dimension of $\mathcal{M}(\widetilde{P})$ is congruent modulo two to $b_1(X) - b_2^+(X) - 1$, so that the dimension is even if and only if $b_1(X) - b_2^+(X)$ is odd.

Lemma 6.7.1 Provided that $b_2^+(X) > 1$, the above definition of $SW(\widetilde{P})$ is independent of the choice of perturbation h and the choice of riemannian metric on X.

Proof. Suppose that we have two riemannian metrics g_0, g_1 on X and two generic perturbations h_0, h_1. Then there is a C^∞ path of metrics g_t joining g_0 to g_1. Choosing a generic path η of perturbations connecting h_0 and h_1 we have a smooth parametrized moduli space

$$\mathcal{M}(\widetilde{P}, \eta) \subset \mathcal{B}^*(\widetilde{P}) \times I.$$

This parametrized moduli space is a compact smooth manifold with boundary. (We need the condition $b_2^+(X) > 1$ in order to ensure that this one-parameter family of moduli spaces contains no reducible solutions.) It is oriented by the choice of orientations of the cohomology spaces of

X and by the orientation on I. Its boundary as an oriented manifold is $\mathcal{M}(\widetilde{P}, h_1) - \mathcal{M}(\widetilde{P}, h_0)$. Thus, the homology classes in $\mathcal{B}^*(\widetilde{P})$ represented by these oriented moduli spaces are the same, and hence

$$\int_{\mathcal{M}(\widetilde{P}, h_1)} \mu^d = \int_{\mathcal{M}(\widetilde{P}, h_0)} \mu^d.$$

This means that the two definitions of the Seiberg-Witten invariant of $\widetilde{P}$ give the same answer. $\square$

Remark 6.7.2 Notice that we have in fact shown that when $b_2^+(X) > 1$ the bordism class of $\mathcal{M}(\widetilde{P}, h)$ in $\mathcal{B}^*(\widetilde{P})$ is an invariant of $\widetilde{P} \to X$.

Theorem 6.7.3 *Let X be a closed, smooth, oriented four-manifold with $b_2^+(X) > 1$. Let $\mathcal{S}(X)$ denote the set of isomorphism classes of $Spin^c$ structures on X. Then the above definition leads to a well-defined invariant, the Seiberg-Witten invariant,*

$$SW : \mathcal{S}(X) \to \mathbf{Z}.$$

This function is zero on all but finitely many elements of $\mathcal{S}(X)$.

Proof. Everything except the last statement was established in the discussion leading up to the statement of the theorem. The last statement is immediate from Theorem 5.2.4. $\square$

6.8 An Involution in the Theory

There is one general relationship between the values of the Seiberg-Witten invariant on various $Spin^c$ structures. There is a complex conjugation involution on $Spin^c$ structures and the value of the Seiberg-Witten invariant on a $Spin^c$ structure differs by a sign from that on its complex conjugate.

Let us begin with some basic variations of complex conjugation. Let V be a real Euclidean space of finite dimension. We have complex conjugation of $Cl(V) \otimes_{\mathbf{R}} \mathbf{C}$, denoted

$$c \mapsto \bar{c}.$$

It induces a map $Spin^c(V) \to Spin^c(V)$ which we call complex conjugation. This map is the identity on $Spin(V) \subset Spin^c(V)$ and which acts by complex conjugation on the determinant.

Now suppose that the dimension of V is congruent to 2 or 4 modulo 8. We denote this dimension by $2d$. Then $Cl(V)$ is isomorphic to a matrix algebra over $\mathbf{H}$, see Section 4 of Chapter 1 of [7]. In this case the irreducible complex module $S_{\mathbf{C}}(V)$ for $Cl(V)$ can be identified with $\mathbf{H}^{2^{d-1}}$ with the Clifford algebra acting by left quaternion matrix multiplication and the complex structure being given by right complex multiplication. There is an isomorphism $\iota \colon S_{\mathbf{C}}(V) \to S_{\mathbf{C}}(V)$ which commutes with Clifford multiplication by $Cl(V)$ and which anti-commutes with the right complex multiplication. This isomorphism is given by right quaternion multiplication by j. For $c \in Cl(V) \otimes_{\mathbf{R}} \mathbf{C}$ and $s \in S_{\mathbf{C}}(V)$ we have

$$\iota(c \cdot s) = \bar{c} \cdot \iota(s).$$

Now let $\widetilde{P} \to X$ be a $Spin^c$-structure for an oriented riemannian four-manifold X. Then letting $P_{SO(n)}$ denote principal bundle associated to the tangent bundle and P_{S^1} be the principal S^1-bundle which it the determinant of $\widetilde{P}$, we have that $\widetilde{P}$ is a double covering of $P_{SO(4)}X \times_X P_{S^1}$. Let $P^*_{S(1)}$ denote the dual (or conjugate) bundle to P_{S^1}. We also have the map

$$(\mathrm{Id}, \iota) \colon P_{SO(4)}X \times_X P^*_{S^1} \to P_{SO(4)}X \times_X P_{S^1}$$

which is the identity on the first factor and which is conjugation of the second factor. Pulling back the double covering induces a $Spin^c$-structure which we denote $-\widetilde{P}$. Its determinant is $P^*_{S^1}$. Of course, there is a map

$$\widetilde{c} \colon (-\widetilde{P}) \to \widetilde{P}$$

covering (Id, c). It is easy to see that $\widetilde{c}$ is a map of principal bundles which is compatible with actions of $Spin^c(n)$ provided that we use complex conjugation on $Spin^c(n)$.

Let $S(\mathbf{R}^n)$ be an irreducible complex module over the Clifford algebra $Cl(\mathbf{R}^n)$. Then there is a map of complex spin bundles.

$$\iota_{\widetilde{P}} \colon (-\widetilde{P}) \times_{Spin^c(n)} S_{\mathbf{C}}(\mathbf{R}^4) \to \widetilde{P} \times_{Spin^c(n)} S_{\mathbf{C}}(\mathbf{R}^4)$$

induced by $\widetilde{c}$ on the principal $Spin^c(n)$ bundles and right multiplication by j on $S_{\mathbf{C}}(\mathbf{R}^4)$. This is a complex anti-linear isomorphism from $S(-\widetilde{P})$ to $S(\widetilde{P})$. The unit $\omega_{\mathbf{C}}$ is in the real Clifford algebra, and thus it is preserved under $\iota_{\widetilde{P}}$. Hence, the isomorphism $\iota_{\widetilde{P}}$ preserves the splittings and induces isomorphisms $\iota^{\pm}_{\widetilde{P}} \colon S^{\pm}(-\widetilde{P}) \to S^{\pm}(\widetilde{P})$. Notice that if c is a section of $Cl(TX) \otimes \mathbf{C}$ and if ψ is a section of $S(-\widetilde{P})$ then we have

$$\iota_{\widetilde{P}}(c \cdot \psi) = \bar{c} \cdot \iota_{\widetilde{P}}(\psi).$$

In particular, if $c \in Cl(TX)$ is an element of the real Clifford algebra, then $\iota_{\widetilde{P}}(c \cdot \psi) = c \cdot \iota_{\widetilde{P}}(\psi)$.

Lemma 6.8.1 *Let A be a connection on P_{S^1} the determinant bundle of $\widetilde{P}$ and let A^* be the conjugate connection on $P^*_{S^1}$. These induce connections ∇_A on $\widetilde{P}$ and ∇_{A^*} on $-\widetilde{P}$, and hence Dirac operators $\partial\!\!\!/_{A^*}$ and $\partial\!\!\!/_A$ on $S^{\pm}(-\widetilde{P})$ and $S^{\pm}(\widetilde{P})$ respectively. For any section ψ of $S(-\widetilde{P})$ we have the following equation*

$$\iota_{\widetilde{P}}(\partial\!\!\!/_{A^*}(\psi)) = \partial\!\!\!/_A(\iota_{\widetilde{P}}(\psi)).$$

Proof.   The map $\iota_{\widetilde{P}} : (-\widetilde{P}) \to \widetilde{P}$ pulls the connection ∇_A back to the connection ∇_{A^*}, and the map $S(-\widetilde{P}) \to S(\widetilde{P})$ commutes with the actions of the real Clifford algebra $Cl(TX)$, the lemma follows immediately. $\square$

Corollary 6.8.2 *The map $\iota_{\widetilde{P}}$ induces complex anti-linear isomorphisms from the spaces of harmonic sections of $S^{\pm}(-\widetilde{P})$ to the corresponding spaces of harmonic sections of $S^{\pm}(\widetilde{P})$.*

Now let us consider what happens to the terms in the curvature equation of Seiberg-Witten. Of course, it is clear that $F^+_{A^*} = -F^+_A$. On the other hand, $q(\psi)$ and $q(\iota_{\widetilde{P}}(\psi))$ are endomorphisms of $S^+(\widetilde{P})$ and $S^+(-\widetilde{P})$ which correspond under $\iota_{\widetilde{P}}$. Since $F^+_{A^*}$ is a purely imaginary two-form, its action via Clifford multiplication on $S^+(-\widetilde{P})$ is carried by $\iota_{\widetilde{P}}$ to the action of $-F^+_{A^*} = F^+_A$ on $S(\widetilde{P})$.

Theorem 6.8.3 *The mapping*

$$(A, \psi) \mapsto (A^*, \iota_{\widetilde{P}}^{-1}(\psi))$$

from the configuration space $\mathcal{C}(\widetilde{P})$ to the configuration space $\mathcal{C}(-\widetilde{P})$ is a diffeomorphism which commutes with the action of the gauge groups when we compare the two gauge groups by complex conjugation. Thus, it induces a homeomorphism of $\mathcal{B}(\widetilde{P}) \to \mathcal{B}(-\widetilde{P})$ which is a diffeomorphism on the open subset of irreducible configurations. This isomorphism of the configurations spaces induces a homeomorphism of the moduli spaces $\mathcal{M}(\widetilde{P}) \to \mathcal{M}(-\widetilde{P})$ which is a linear isomorphism on the Zariski tangent space at each irreducible solution. Furthermore, if h is a self-dual two-form then the identification of configuration spaces induces a homeomorphism

$$\mathcal{M}(P, h) \to \mathcal{M}(-\widetilde{P}, -h).$$

In particular, one of these moduli spaces is smooth if and only if the other is and in this case the map is an diffeomorphism.

Now let us translate this into a statement about the value of the Seiberg-Witten invariants $SW(-\widetilde{P})$ and $SW(\widetilde{P})$.

Corollary 6.8.4 *Let X be a closed, oriented four-manifold. Suppose that $b_2^+(X) > 1$, and fix orientations on $H^1(X;\mathbf{R})$ and $H_+^2(X;\mathbf{R})$ inducing orientations on all the moduli spaces, and hence determining the sign of the Seiberg-Witten invariants for every $Spin^c$ structure on X. We have*

$$SW(-\widetilde{P}) = (-1)^{\epsilon(X)} SW(\widetilde{P})$$

where

$$\epsilon(X) = (1 + b_2^+(X) - b_1(X))/2.$$

Proof. We need only compute the effect of the involution on the orientation of the moduli spaces and the effect of the involution on the class μ. First of all, complex conjugation on the gauge group induces a conjugate isomorphism between the universal circle bundles over $\mathcal{B}^*(\widetilde{P})$ and $\mathcal{B}^*(-\widetilde{P})$. Hence, the identification of the configuration spaces send μ to $-\mu$.

Now let us consider the effect of the involution on the determinant line bundles over $\mathcal{B}^*(\widetilde{P})$ and $\mathcal{B}^*(-\widetilde{P})$. It suffices to consider the effect on the determinant line bundle of the homotopic family $(d^+ + 2d^* + \partial\!\!\!/_A)$. The map sends i to $-i$ and hence acts by -1 on the groups $H^0(X;i\mathbf{R})$, $H^1(X;i\mathbf{R})$, and $H_+^2(X;i\mathbf{R})$. Also, it is a complex anti-linear isomorphism on the harmonic plus and minus spinors. Thus, the total effect of this isomorphism on the determinant line bundles and hence on the orientations of the moduli spaces is $(-1)^\delta$ where $\delta = 1 + b_1(X) + b_2^+(X) + \mathrm{ind}_{\mathbf{C}}(\partial\!\!\!/_A)$.

Let d be the formal dimension of the moduli space $\mathcal{M}(\widetilde{P})$. Then the isomorphism of configuration spaces sends $\mu^{d/2}$ by $(-1)^{d/2}$.

These two facts together imply that the effect of the involution on the Seiberg-Witten invariant is to multiply the value by

$$(-1)^{\delta + d/2}.$$

Of course, by the index theorem we have

$$d/2 = (b_1(X) - 1 - b_2^+(X))/2 + \mathrm{ind}_{\mathbf{C}}(\partial\!\!\!/).$$

Putting all this together, we see that

$$SW(-\widetilde{P}) = (-1)^{\epsilon(X)} SW(\widetilde{P}).$$

$\square$

6.9   The Case when $b_2^+(X) = 1$

Here, we assume that X is a closed, oriented smooth four-manifold with $b_2^+(X) = 1$. We choose orientations for $H^1(X;\mathbf{R})$ and $H_+^2(X;\mathbf{R})$. We proceed in much the same manner as above with one exception. We must exclude the codimension-one submanifold of metrics for which there are reducible solutions of the Seiberg-Witten equations.

Lemma 6.9.1 *Let $b_2^+(X) = 1$ and let g be a metric for X. Then there is a reducible solution to the Seiberg-Witten equations for a $Spin^c$ structure $\widetilde{P}$ if and only if the cohomology class represented by a non-trivial g self-dual harmonic two-form is orthogonal to $c_1(\mathcal{L})$ under the cup product on cohomology.*

Proof.   There is a reducible solution to the Seiberg-Witten equations if and only if there is an ASD connection on $\mathcal{L}$. The curvature of such a connection would then be an g ASD harmonic two-form. Such a connection exists if and only if $c_1(\mathcal{L})$ is orthogonal under cup product to the space of g self-dual harmonic two-forms. But this space is one-dimensional. The result follows.   □

Now suppose that an non-zero g self-dual harmonic two-form is not orthogonal to $c_1(\mathcal{L})$. Then there are no reducible solutions to the Seiberg-Witten equations. For a sufficiently small perturbation h there will be no reducible solutions for the perturbed equations either. Thus, for a sufficiently small generic h the moduli space $\mathcal{M}(\widetilde{P}, h)$ will be a compact smooth manifold. It represents a cycle in $\mathcal{B}^*(\widetilde{P})$ and, provided that its dimension is even (say $2d$), we can define the Seiberg-Witten invariant of (X, g) on $\widetilde{P}$ to be

$$SW_g(\widetilde{P}) = \int_{\mathcal{M}(\widetilde{P}, h)} \mu^d,$$

where, as in the case when $b_2^+(X) > 1$ we use the cohomology orientations for X to orient the moduli space. One sees easily that for all sufficiently small perturbations this integral gives the same result, so that $SW_g(\widetilde{P})$ is well-defined.

In this context, we have defined an invariant not just of X but also of the component of the space of riemannian metrics on X with the property that the self-dual harmonic two-form of the metric is not orthogonal to $c_1(\mathcal{L})$. Next, we shall see that, at least in the case when $H_1(X;\mathbf{Z}) = 0$, the value of the invariant depends only on the sign of the intersection of a non-zero self-dual two-form giving the orientation of $H_+^2(X; \mathbf{R})$ with $c_1(\mathcal{L})$. This means that there are only two values, depending on the metric, for the Seiberg-Witten invariant of a given $Spin^c$ structure. The difference of these invariants is a interesting feature of the theory.

The Wall-Crossing Formula   In this paragraph we concentrate on the case when $b_2^+(X) = 1$. Fix a $Spin^c$ bundle $\tilde{P} \to X$. As we have seen, if we have a smooth one-parameter family of metrics g_t with the property that for no member of the family has a reducible solution, then for generic family $\eta = \{h_t\}_{0 \le t \le 1}$ of sufficiently small self-dual two-forms, the parametrized moduli space $\mathcal{M}(\tilde{P}, \eta)$ is a closed, oriented smooth manifold and hence the Seiberg-Witten invariants of $\tilde{P}$ computed using the metrics g_0 and g_1 are equal. This argument does not work if there are reducible solutions to the Seiberg-Witten equations for some g_t in the family. This happens when the line of self-dual harmonic two-forms for g_t is orthogonal under the intersection pairing to $c_1(\mathcal{L})$.

The map from metrics to lines in $H^2(X; \mathbf{R})$ which associates to each metric the line of self-dual harmonic two-forms is a submersion. Fix an orientation of $H_+^2(X; \mathbf{R})$. Then for each metric g_t there is a unique g_t-self-dual harmonic two-form of norm one which lies in the positive component of $H_+^2(X; \mathbf{R})$ as measured by the given orientation. We denote this form by $\omega^+(g_t)$.

Theorem 6.9.2 *Let X be a closed, oriented four-manifold with*

$$H_1(X; \mathbf{Z}) = 0$$

and $b_2^+(X) = 1$. Fix an orientation for $H_+^2 X; \mathbf{R})$ and let $\tilde{P}$ be a $Spin^c$ structure on X such that $c_1(\mathcal{L}) \ne 0$. Let $\mathcal{R}$ be the space of riemannian metrics on X. Let $\mathcal{R}_+$ be the subset of metrics g such that $\omega^+(g) \cdot c_1(\mathcal{L}) > 0$ and let $\mathcal{R}_-$ be the subset of metrics where this same intersection is negative. Then for each metric $g \in \mathcal{R}' = \mathcal{R}_+ \coprod \mathcal{R}_-$ the Seiberg-Witten invariant $SW_g(\tilde{P})$ is defined. Furthermore, the function on $\mathcal{R}'$ which assigns to each metric g in this subset the value of the Seiberg-Witten invariant $SW_g(\tilde{P})$ is constant on each of $\mathcal{R}_\pm$. We denote the common value on $\mathcal{R}_\pm$ by $SW_\pm(\tilde{P})$. Also, provided that the formal dimension of the moduli space $\mathcal{M}(\tilde{P})$ is non-negative and even, we have

$$SW_+(\tilde{P}) = SW_-(\tilde{P}) - (-1)^{d/2}.$$

The rest of this subsection is devoted to the proof of this theorem. We begin with an elementary lemma.

Lemma 6.9.3 *Let X be a closed, oriented riemannian 4-manifold. Suppose that $\widetilde{P} \to X$ is a $Spin^c$ structure with the property that the Dirac operator from sections of $S^+(\widetilde{P})$ to sections of $S^-(\widetilde{P})$ has non-negative index. For any connection A on $\mathcal{L}$ there is an open dense subset of $a \in \Omega^1(X; i\mathbf{R})$ such that the operator $D_{A,a}$ from sections of $S^+(\widetilde{P})$ to sections of $S^-(\widetilde{P})$ given by*

$$D_{A,a}(\psi) = \partial\!\!\!/_A(\psi) + a \cdot \psi$$

is surjective.

Proof.   It follows from the computation in the proof of Lemma 6.2.1 that for a generic L^2_2 one-form a that the operator $D_{A,a}$ is surjective. Of course, it is clear that the subset of a for which $D_{A,a}$ is surjective is an open subset. $\qquad\qquad\qquad\qquad\qquad\qquad\qquad\qquad\qquad\qquad\qquad\qquad$ $\square$

Now suppose that X is as in the statement of the theorem. Choose an orientation for $H^2_+(X; \mathbf{R})$. By a standard transversality argument, Theorem 6.9.2 is a consequence of the following local computation.

Proposition 6.9.4 *Suppose also that $\{g_t\}_{0 \leq t \leq 1}$ is a smooth path of metrics on X with the property that the only metric in the family with reducible connections is $g_{1/2}$ and furthermore, that the intersection of the self-dual two-forms $\omega(g_t)$ with $c_1(\mathcal{L})$ defines a function which is transverse to zero at $t = 1/2$, passing from negative to positive. Provided that the formal dimension d of the moduli space $\mathcal{M}(\widetilde{P})$ is at least zero and is even, we have*

$$SW_{g_1}(X) - SW_{g_0}(X) = -(-1)^{d/2}.$$

Proof.   The assumptions on $H^*(X; \mathbf{R})$ imply that the formal dimension d of $\mathcal{M}(\widetilde{P})$ is equal to $2\,\mathrm{ind}_{\mathbf{C}}(\partial\!\!\!/_{A_0}) - 2$. Hence, the assumption that this dimension is non-negative means that $\mathrm{ind}_{\mathbf{C}}(\partial\!\!\!/_{A_0}) \geq 1$. Denote by γ the path of metrics. The moduli space of solutions

$$\mathcal{M}_\gamma(\widetilde{P}) \subset \mathcal{B}(\widetilde{P}) \times I$$

is a compact submanifold. It has one reducible solution, say $([A_0, 0], t)$. Choose a generic $a \in \Omega^1(X; i\mathbf{R})$. According to Lemma 6.9.3, the operator $D_{A_0,a}$ given by

$$D_{A_0,a}(\psi) = \partial\!\!\!/_{A_0}(\psi) + a \cdot \psi$$

has no cokernel.

We now define the perturbed moduli space of all solutions of

$$F_A^+ = q(\psi)$$
$$\partial_A(\psi) + a \cdot (\psi) = 0.$$

Notice that the perturbed equations have exactly the same set of reducible solutions as the original equations. Since a and its first derivative are pointwise bounded and since $D_{A_0,a}$ is an elliptic operator, the arguments from the previous sections imply that the parametrized moduli space for these equations and the given family of metrics $\mathcal{M}'(\tilde{P}, \gamma)$ is a compact moduli space with only one reducible solution. Since cokernel of $D_{A_0,a}$ is trivial and since the partial derivative in the t direction maps onto $H_+^2(X; i\mathbf{R})$ in an orientation-preserving fashion, it follows that at the reducible solution the second cohomology of the elliptic complex for the parametrized moduli space is trivial. A neighborhood of this solution is then isomorphic to $\mathbf{C}^{(d/2)+1}$ modulo the action of the stabilizer of A_0. Thus, this neighborhood in the moduli space is homeomorphic to the cone over a complex projective space.

Let us compute the orientation at any point of this neighborhood different from the singularity. The kernel of our operator $(D_{A_0,a}, d^*)$ from $\mathcal{C}(\tilde{P}) \times I$ to $\Omega_+^2(X; i\mathbf{R}) \oplus \Omega^0(X; i\mathbf{R})$ is the space of harmonic spinors with the orientation induced from the complex structure.

Claim 6.9.5 *The cokernel is* $H^0(X; i\mathbf{R})$ *with the canonical orientation.*

Proof. The fact that the self-dual harmonic two-forms $\omega^+(g_t)$ give a path in cohomology which crosses the wall perpendicular to $c_1(\mathcal{L})$ transversely from negative to positive, means that the inner products

$$\omega^+(g_t) \cdot c_1(\mathcal{L})$$

form a real-valued function which is transverse to zero crossing from negative to positive as t increases. This means that the harmonic projections of $F_{A_0}^{+\prime}$ cross from negative multiples of $(2\pi/i)\omega^+(g_t)$ to positive multiples as t increases. That is to say, in the given orientation of $H_+^2(X; i\mathbf{R})$ this path of intersections is transverse to zero crossing from the positive side to the negative side as t increases. We have just established that image under the differential of the tangent vector ∂_t in the domain points in the negative direction in $H_+^2(Xi; \mathbf{R})$. Now we divide out the orientation of $H_+^2(X; i\mathbf{R}) \oplus H^0(X; i\mathbf{R})$ by this image vector. Since in dividing out we put the vector last, it follows that the quotient orientation on the range is the usual orientation on $H^0(X; i\mathbf{R})$. This completes the proof of the claim. $\square$

Of course, the action of the stabilizer, S^1, of the reducible solution on the harmonic spinors is the opposite of the usual complex action. This means that the orientation of the quotient space is the opposite to the oriented quotient of a complex linear space by the usual complex action. Viewing this quotient as the cone of a complex projective space, the orientation of the moduli space (away from the singular point) is the opposite of the complex orientation on the complex projective space followed by the tangent vector pointing away from the singularity in the cone direction. In particular, cutting out a neighborhood of the singularity, the boundary of the remaining manifold is the complex orientation on the complex projective space. Of course, μ is induced by the circle action and hence is the negative of the usual generator of the second cohomology. It follows that the evaluation of $\mu^{d/2}$ on this boundary is $(-1)^{d/2}$.

We choose a generic path $\eta = h(t) \in \Omega^2_+(X; \mathbf{R})$ so that the parametrized moduli space $\mathcal{M}(P, \eta)$ is a smooth manifold except possibly at the reducible points. If $h(t)$ and $\partial_t h(t)$ are sufficiently small, then there will be only one reducible and this will be a point where the function

$$t \mapsto \left(\omega^+(g_t) - \pi(h(t))\right) \cdot c_1(\mathcal{L})$$

is transverse to zero. (Here π denotes the L^2-projection on to the self-dual harmonic forms.) Furthermore, since the second cohomology of the elliptic complex at the original reducible solution is trivial, this is also true for the perturbed equations, provided that h is sufficiently small. Thus, the complement in the parametrized moduli space of a neighborhood of the reducible solution is a compact oriented manifold with boundary. Its boundary as an oriented manifold is

$$\mathcal{M}(\widetilde{P}, h_1) - \mathcal{M}(\widetilde{P}, h_0) + \mathbf{C}P^{d/2}.$$

Thus, evaluating $\mu^{d/2}$ on this boundary gives zero and hence we conclude that

$$0 = SW_{g_1}(\widetilde{P}) - SW_{g_0}(\widetilde{P}) + (-1)^{d/2}$$

or equivalently that

$$SW_{g_1}(\widetilde{P}) - SW_{g_0}(\widetilde{P}) = -(-1)^{d/2}.$$

This completes the proof of the proposition. $\square$

As we remarked above, the theorem is an easy consequence of the proposition.

Chapter 7

Invariants of Kähler Surfaces

In order to explicitly evaluate the Seiberg-Witten invariant of a compact Kähler surface we begin by rewriting the Seiberg-Witten equations in terms of the complex geometry. From this description it will be a fairly easy matter to completely solve the equations over such a four-manifold in terms of standard holomorphic objects.

7.1   The Equations over a Kähler Manifold

In this section we shall give an explicit description of the Seiberg-Witten moduli space when the base riemannian four-manifold is a compact Kähler surface. Recall from Chapter 3 that an orthogonal almost complex structure $J : TX \to TX$ on a riemannian four-manifold induces a $Spin^c$ structure $\widetilde{P}_J \to X$ whose determinant line bundle is K_X^{-1}, the inverse of the canonical line bundle of the almost complex structure. The spin bundles are given by

$$S^+(\widetilde{P}_J) = \Lambda^0(X; \mathbf{C}) \oplus \Lambda^{0,2}(X; \mathbf{C})$$

$$S^-(\widetilde{P}_J) = \Lambda^{0,1}(X; \mathbf{C}).$$

Clifford multiplication by a one-form $a \in \Omega^1(X; \mathbf{C})$ is given by the sum of wedge product and minus contraction with $\sqrt{2}\pi^{0,1}(a) \in \Omega^{0,1}(X; \mathbf{C})$ of a. Furthermore, if the almost complex structure is in fact a complex structure for which the riemannian metric is a Kähler metric, then the Dirac operator on plus spinors associated to this $Spin^c$ structure and the natural holomorphic, hermitian connection on K_X^{-1} is

$$\sqrt{2}\left(\overline{\partial} + \overline{\partial}^*\right) : \Omega^0(X; \mathbf{C}) \oplus \Omega^{0,2}(X; \mathbf{C}) \to \Omega^{0,1}(X; \mathbf{C}).$$

Any other $Spin^c$ structure $\widetilde{P}$ differs from $\widetilde{P}_J$ by tensoring with some $U(1)$-bundle $Q \to X$. Let $\mathcal{L}_0$ be the complex line bundle associated to Q. Then the spin bundles for $\widetilde{P}$ are given by

$$S^+(\widetilde{P}) = S^+(\widetilde{P}_J) \otimes \mathcal{L}_0 = \Lambda^0(X; \mathcal{L}_0) \oplus \Lambda^{0,2}(X; \mathcal{L}_0)$$

$$S^-(\widetilde{P}) = S^-(\widetilde{P}_J) \otimes \mathcal{L}_0 = \Lambda^{0,1}(X; \mathcal{L}_0).$$

As before, Clifford multiplication by $a \in \Omega^1(X; \mathbf{C})$ is given by the sum of wedge product and minus contraction with $\sqrt{2}\pi^{0,1}(a) \in \Omega^{0,1}(X; \mathbf{C})$. Furthermore, the determinant of $\widetilde{P}$ is identified with $K_X^{-1} \otimes \mathcal{L}_0^2$, or put another way, $\mathcal{L}_0$ is a square root of $K_X \otimes \mathcal{L}$, where as usual $\mathcal{L} = \det(\widetilde{P})$. A $U(1)$-connection A on $\mathcal{L}$ is equivalent to a unitary connection A_0 on $\mathcal{L}_0$, the equivalence being given by

$$(A_0)^2 = A_{K_X} \otimes A$$

where A_{K_X} is the holomorphic, hermitian connection on K_X. The Dirac operator associated to the connection A on $\mathcal{L}$ is

$$\sqrt{2}\left(\overline{\partial}_{A_0} + \overline{\partial}_{A_0}^*\right) : \Omega^0(X; \mathcal{L}_0) \oplus \Omega^{0,2}(X; \mathcal{L}_0) \to \Omega^{0,1}(X; \mathcal{L}_0),$$

the operator obtained by coupling $\sqrt{2}\left(\overline{\partial} + \overline{\partial}^*\right)$ with the covariant derivative ∇_{A_0} on $\mathcal{L}_0$. Notice that the notation is not meant to suggest that $\overline{\partial}_{A_0}$ determines a holomorphic structure, for in general, this need not be true.

We now have a good description of the Dirac equation: A spinor field ψ, a section of $S^+(\widetilde{P})$ has two components

$$\psi = (\alpha, \beta) \in \Omega^0(X; \mathcal{L}_0) \oplus \Omega^{0,2}(X; \mathcal{L}_0)$$

and the Dirac equation is

$$\sqrt{2}\left(\overline{\partial}_{A_0}(\alpha) + \overline{\partial}_{A_0}^*(\beta)\right) = 0.$$

Let us write out the curvature equation with respect to the Kähler geometry. Let ω be the Kähler form. It is a nowhere zero, self-dual, real two-form of type $(1,1)$. The complex-valued self-dual-two forms on a Kähler manifold split as

$$\Omega^0(X;\mathbf{C}) \cdot \omega \oplus \left(\Omega^{2,0}(X;\mathbf{C}) \oplus \Omega^{0,2}(X;\mathbf{C}) \right).$$

The purely imaginary self-dual two-forms are then

$$\Omega^0(X;i\mathbf{R}) \cdot \omega \oplus \{\mu - \overline{\mu}|\mu \in \Omega^{0,2}(X;\mathbf{C})\}.$$

Hence, the self-dual part of the curvature of the unitary connection A on $\mathcal{L}$ can be written as

$$F_A^+ = if\omega + \mu - \overline{\mu}$$

for some real-valued function f on X and some complex-valued $(0,2)$-form μ on X.

Let us examine the endomorphism of $S^+(\widetilde{P})$ induced by such a self-dual two-form. We take local holomorphic coordinates (z_1, z_2) centered at a point so that the Kähler metric is standard to second order at the point. Let us compute the action of $d\overline{z}_1 \wedge d\overline{z}_2$ on $S^+(\widetilde{P})$ by Clifford multiplication. Since $d\overline{z}_1$ and $d\overline{z}_2$ are orthogonal, we see that the element in the Clifford algebra corresponding to this wedge product is simply the Clifford product $d\overline{z}_1 \cdot d\overline{z}_2$. Similarly, the element in the Clifford algebra corresponding to $dz_1 \wedge dz_2$ is simply the Clifford product $dz_1 \cdot dz_2$. Since $\pi^{0,1}(dz_i) = 0$, it follows immediately that the action of $dz_1 \wedge dz_2$ on the bundle of plus spinors is trivial. Let us compute the action of $d\overline{z}_j$ on $S^+(P)$. Of course, $\sqrt{2}\pi^{0,1}(d\overline{z}_j) = \sqrt{2}d\overline{z}_j$. Since contraction with $d\overline{z}_1$ following multiplication by $d\overline{z}_2$ is trivial on $\Omega^0(X;\mathcal{L})$ it follows that the action of $d\overline{z}_1 \wedge d\overline{z}_2$ on $\Omega^0(X;\mathcal{L}_0)$ is given by wedge product with $2d\overline{z}_1 \wedge d\overline{z}_2$. More generally, the action of any $(0,2)$-form μ by Clifford multiplication on $\Omega^0(X;\mathcal{L}_0)$ is simply given by wedge product with 2μ. In the same vein we see that the action of a $(0,2)$-form by Clifford multiplication on $\Omega^{0,2}(X;\mathcal{L}_0)$ is given by contraction by twice this element. A straightforward computation shows that the contraction by μ on a $(0,2)$-form λ is $*(\mu \wedge \overline{\lambda})$. (Here, we are using the hermitian extension of $*$ which is given by $*f\, dvol = \overline{f}$.)

Let us consider the action of the Kähler form ω. In the given local coordinates we write

$$\omega = dx_1 \wedge dy_1 + dx_2 \wedge dy_2.$$

Since dx_i and dy_i are orthogonal, the element in the Clifford algebra corresponding to this element is

$$dx_1 dy_1 + dx_2 dy_2.$$

(Notice that this element is not equal to $i(dz_1 d\bar{z}_1 + dz_2 d\bar{z}_2)/2$. The problem is that dz_1 and $d\bar{z}_1$ are not orthogonal.) If we begin in $\Omega^0(X; \mathcal{L}_0)$, then the first operation is multiplication by $\frac{i}{\sqrt{2}}(d\bar{z}_1 + d\bar{z}_2)$ followed by minus the contraction by $\frac{1}{\sqrt{2}}(d\bar{z}_1 + d\bar{z}_2)$. Direct computation shows that the result is multiplication by $-2i$. If we begin in $\Omega^{0,2}(X; \mathcal{L}_0)$, then we must first take minus the contraction and then wedge. Since contraction is anti-linear in the first variable, we see that this composition is multiplication by $2i$.

It follows from these computations that if

$$F_A^+ = if\omega + \mu - \bar{\mu}$$

for some real-valued function f and some complex-valued $(0,2)$-form μ then the endomorphism induced by Clifford multiplication by F_A^+ is

$$\begin{pmatrix} 2f & *2\left(\mu \wedge \overline{(\cdot)}\right) \\ 2\mu \wedge (\cdot) & -2f \end{pmatrix}.$$

On the other hand, the matrix representative for $\psi \otimes \bar{\psi}$ is

$$\begin{pmatrix} \alpha \\ \beta \end{pmatrix} \cdot \begin{pmatrix} \bar{\alpha} & \bar{\beta} \end{pmatrix} = \begin{pmatrix} |\alpha|^2 & \alpha\bar{\beta} \\ \bar{\alpha}\beta & |\beta|^2 \end{pmatrix}$$

Of course $|\psi|^2 = |\alpha|^2 + |\beta|^2$. This means that the matrix representative for $q(\psi)$ is

$$\begin{pmatrix} \frac{|\alpha|^2 - |\beta|^2}{2} & \alpha\bar{\beta} \\ \bar{\alpha}\beta & \frac{|\beta|^2 - |\alpha|^2}{2} \end{pmatrix}$$

Thus, we see that the curvature equation is equivalent to the following equations:

$$(F_A^+)^{1,1} = \frac{i}{4}\left(|\alpha|^2 - |\beta|^2\right)\omega \tag{7.1}$$

$$F_A^{0,2} = \frac{\bar{\alpha}\beta}{2}. \tag{7.2}$$

7.2   Holomorphic Description of the Moduli Space

Now let us solve the equations that we derived in the last section in terms of the holomorphic geometry. Let us fix a connected Kähler surface X and a *Spin*^{c} structure $\tilde{P}$ whose determinant line bundle is $\mathcal{L}$ and so that the associated square root of $K_X \otimes \mathcal{L}$ is $\mathcal{L}_0$. We define the *degree* of $\mathcal{L}$ by

$$\deg(\mathcal{L}) = \int_X c_1(\mathcal{L}) \wedge \omega.$$

Lemma 7.2.1 *Fix a Spin^c-structure $\tilde{P}$ for a Kähler surface X. Let $\mathcal{L}$ be the determinant line bundle for $\tilde{P}$ and set $\mathcal{L}_0$ equal to the square root of $K_X \otimes \mathcal{L}$ determined by $\tilde{P}$. Let (A, ψ) be a solution to the Seiberg-Witten equations for $\tilde{P}$. As in the last section we write $\psi = (\alpha, \beta)$ with $\alpha \in \Omega^0(X; \mathcal{L}_0)$ and $\beta \in \Omega^{0,2}(X; \mathcal{L}_0)$. If the degree of $\mathcal{L}$ is ≤ 0, then we have $\beta = 0$, and if the degree of $\mathcal{L}$ is ≥ 0 then $\alpha = 0$. Furthermore, A induces a holomorphic structure on $\mathcal{L}$. With respect to the induced holomorphic structure on $\mathcal{L}_0$, the section α is holomorphic and $\bar{\beta}$ is a holomorphic section of $K_X \otimes \mathcal{L}_0^{-1}$.*

Proof. Let us begin by proving that $\bar{\alpha}\beta$ is zero. We have the harmonic spinor equation

$$\sqrt{2}\left(\bar{\partial}_{A_0}(\alpha) + \bar{\partial}^*_{A_0}(\beta)\right) = 0. \tag{7.3}$$

Applying $\sqrt{2}^{-1}\bar{\partial}_{A_0}$ to this equation we get

$$\bar{\partial}_{A_0}\bar{\partial}_{A_0}(\alpha) + \bar{\partial}_{A_0}\bar{\partial}^*_{A_0}(\beta) = 0. \tag{7.4}$$

Of course, $\bar{\partial}_{A_0}\bar{\partial}_{A_0}(\alpha) = F^{0,2}_{A_0} \cdot \alpha$. Since A_0 is the connection on $\sqrt{K_X \otimes \mathcal{L}}$ which is the square root of the tensor product of a holomorphic connection on K_X and the connection A on $\mathcal{L}$, it is clear from Equation 7.2 that

$$F^{0,2}_{A_0} = \frac{1}{2}F^{0,2}_A = \frac{1}{4}\bar{\alpha}\beta.$$

Plugging this into Equation 7.4 gives

$$\frac{1}{4}|\alpha|^2\beta + \bar{\partial}_{A_0}\bar{\partial}^*_{A_0}(\beta) = 0.$$

Taking the L^2-inner product with β yields

$$\int_X \frac{1}{2}|\alpha|^2|\beta|^2 dvol + \|\bar{\partial}^*_{A_0}(\beta)\|^2_{L^2} = 0.$$

Since each of these terms is non-negative, it follows that they both vanish. Of course, $|\bar{\alpha}\beta|^2 = |\alpha|^2|\beta|^2$, so that we conclude that $\bar{\alpha}\beta = 0$.

This means that $F^{0,2}_A = 0$, and hence that A is holomorphic connection. It follows that A_0 is also a holomorphic connection. We also see that $\bar{\partial}^*_{A_0}(\beta) = 0$. This implies that β is an anti-holomorphic section or equivalently that $\bar{\beta}$ is a holomorphic two-form with values in $\bar{\mathcal{L}}_0 = \mathcal{L}_0^{-1}$. From Equation 7.3 it now follows that α is a holomorphic section of $\mathcal{L}_0$. In particular, since X is connected, if either α or β vanishes on an open subset of X, then it vanishes identically on X. Thus, we see that one of α and β is identically zero since their product is identically zero.

All that remains is to show that the sign of the degree of $\mathcal{L}$ determines which of α and β is zero. We have

$$(F_A^+)^{1,1} = \frac{i}{4}(|\alpha|^2 - |\beta|^2)\omega.$$

Thus, we see that

$$\deg(\mathcal{L}) = \int_X c_1(\mathcal{L}) \wedge \omega = \frac{1}{8\pi} \int_X \left(|\beta|^2 - |\alpha|^2\right) dvol.$$

Since at least one of α and β is zero, we see that if the degree of $\mathcal{L}$ is non-negative, then $\alpha = 0$ and if the degree of $\mathcal{L}$ is non-positive, then $\beta = 0$. $\square$

Corollary 7.2.2 *If the degree of $\mathcal{L}$ is non-positive, then any solution to the Seiberg-Witten equations consists of a holomorphic, hermitian connection A on $\mathcal{L}$ and a holomorphic section α of $\mathcal{L}_0$ with*

$$(F_A^+)^{1,1} = \frac{i}{4}|\alpha|^2\omega.$$

Two such pairs (A, α) and (A', α') determine the same point in the moduli space if and only if there is a holomorphic, hermitian isomorphism between the holomorphic structures $\mathcal{L}_A$ and $\mathcal{L}_{A'}$ such that the induced holomorphic isomorphism $\mathcal{L}_0$ to $\mathcal{L}_0$ carries α to α'.

Proof.   We have seen that the conditions on (A, α) stated in the corollary are equivalent to the fact that this pair yields a solution to the Seiberg-Witten equations and that any solution to the equations arises in this way provided that the degree of $\mathcal{L}$ is non-positive. The uniqueness statement is clear. $\square$

There is a similar result in the case that the degree of $\mathcal{L}$ is non-negative. There are two ways to establish the other case. One is to directly apply the analogous arguments. The other is to use the involution on $Spin^c$-structures described in Section 6.8 which inverts the line bundle. Let us consider this involution. Since it sends $\mathcal{L}$ to its inverse, it sends $\mathcal{L}_0$ to $K_X \otimes \mathcal{L}_0^{-1}$. We have an identification of $S_{\mathbf{C}}^+(\mathbf{R}^4)$ with $\mathbf{H}$. The subspaces $\mathbf{C}$ and $j\mathbf{C}$ are invariant under the action of $U(2)$. These factors yield $\Lambda^0(X; \mathcal{L}_0)$ and $\Lambda^{0,2}(X; \mathcal{L}_0)$ in $S_{\mathbf{C}}^+(\widetilde{P})$. Thus, the right multiplication action of j on $S_{\mathbf{C}}^+(\mathbf{R}^4)$ induces a bundle isomorphism of $S_{\mathbf{C}}^+(\widetilde{P})$ to $S_{\mathbf{C}}^+(-\widetilde{P})$ which takes (α, β) to $(-\overline{\beta}, \overline{\alpha})$. Using this, one deduces the follwing result from Corollary 7.2.2.

Corollary 7.2.3 *If the degree of $\mathcal{L}$ is non-negative, then a solution to the Seiberg-Witten equations consists of a holomorphic, hermitian connection A on $\mathcal{L}$ and a holomorphic section $\overline{\beta}$ of $K_X \otimes \mathcal{L}_0^{-1}$ with*

$$(F_A^+)^{1,1} = -\frac{i}{4}|\beta|^2\omega.$$

Two such pairs (A, β) and (A', β') determine the same point in the moduli space if and only if there is a holomorphic, hermitian isomorphism between the holomorphic structures $\mathcal{L}_A$ and $\mathcal{L}_{A'}$ such that the induced holomorphic isomorphism $K_X \otimes \mathcal{L}_0^{-1}$ to $K_X \otimes \mathcal{L}_0^{-1}$ carries $\overline{\beta}$ to $\overline{\beta}'$.

It remains to deal with the last part of the curvature equation in order to show that when the degree of $\mathcal{L}$ is negative then any holomorphic structure on $\mathcal{L}$ and any non-zero holomorphic section α of $\mathcal{L}_0$ determine a point in the moduli space.

Lemma 7.2.4 *Suppose that the degree of $\mathcal{L}$ is negative, suppose that A is a hermitian, holomorphic connection on $\mathcal{L}$, and suppose that α is a non-zero holomorphic section of $\mathcal{L}_0$ (with respect to the holomorphic structure defined by A_0). Then there exists another hermitian structure h' on $\mathcal{L}$ such that for the connection A' which is hermitian with respect to h' and which defines the same holomorphic structure on $\mathcal{L}$ as A does we have*

$$F_{A'}^{1,1} = \frac{i}{4}(|\alpha|_{h'})^2\omega \tag{7.5}$$

where $(|\alpha|_{h'})^2$ means the norm measured with respect to the hermitian structure on $\mathcal{L}_0$ determined by h'.

Proof. Let us denote by h the given hermitian inner product on $\mathcal{L}$. A new hermitian structure h' on $\mathcal{L}$ is given by $\exp(\lambda)h$ for some smooth real-valued function λ. Of course $(|\alpha|_{h'})^2 = \exp(\lambda)|\alpha|^2$. The curvature of the holomorphic connection A' which is hermitian with respect to h' is given by

$$F_{A'} = F_A + \overline{\partial}\partial(\lambda).$$

Thus, the equation that we need to solve for λ is

$$F_A^+ + \left(\overline{\partial}\partial(\lambda)\right)^+ = \frac{i}{4}\exp(\lambda)|\alpha|^2\omega,$$

or equivalently

$$F_A \wedge \omega + \overline{\partial}\partial(\lambda) \wedge \omega = \frac{i}{4}\exp(\lambda)|\alpha|^2 \omega \wedge \omega. \qquad (7.6)$$

Of course, since the metric is Kähler we have

$$\overline{\partial}\partial\lambda \wedge \omega = \frac{2}{i}\left(-\frac{\partial^2\lambda}{\partial z_1 \overline{\partial} z_1} - \frac{\partial^2\lambda}{\partial z_2 \overline{\partial} z_2}\right)\mathrm{dvol} = \frac{2}{i}\Delta(\lambda)\mathrm{dvol}.$$

Also, since the degree of $\mathcal{L}$ is negative, we have

$$\int_X (iF_A \wedge \omega) < 0.$$

Thus, we can rewrite Equation 7.6 as

$$\Delta(\lambda) + \frac{|\alpha|^2}{4}\exp(\lambda) + C = 0$$

where C is the smooth function with $C\mathrm{dvol} = \frac{1}{2}(F_A \wedge \omega)$. Because of the degree condition on $\mathcal{L}$

$$\int_X C\mathrm{dvol} < 0.$$

According to [5], such an equation has a unique solution $\lambda\colon X \to \mathbf{R}$. $\square$

Since any two hermitian structures on $\mathcal{L}$ are isomorphic by a complex linear bundle automorphism of $\mathcal{L}$, this leads us then to our final description of the moduli space.

Corollary 7.2.5 *Let $\widetilde{P}$ be a Spin^c structure on a Kähler surface X. Suppose that the degree of the determinant $\mathcal{L}$ is negative. Fix a pair $(\overline{\partial}_{\mathcal{L}}, \alpha_0)$, where $\overline{\partial}_{\mathcal{L}}$ is a holomorphic structure on $\mathcal{L}$ and α_0 is a non-zero holomorphic section of $\mathcal{L}_0 = \sqrt{K_X \otimes \mathcal{L}}$. Then there is a solution (A, α) to the Seiberg-Witten equations as in Corollary 7.2.2 with the following properties:*

- *A determines a holomorphic structure on $\mathcal{L}$ which is isomorphic to the holomorphic structure $\overline{\partial}_{\mathcal{L}}$, and*

- *there is a holomorphic isomorphism from the structure determined by A to $\overline{\partial}_{\mathcal{L}}$ which sends α to α_0.*

Such a solution (A, α) is unique up to gauge equivalence. In this way any pair $(\overline{\partial}_{\mathcal{L}}, \alpha_0)$ as above determines a point in the moduli space $\mathcal{M}(\widetilde{P})$. All points of $\mathcal{M}(\widetilde{P})$ arise in this way. Two pairs $(\overline{\partial}_{\mathcal{L}}, \alpha_0)$ and $(\overline{\partial}'_{\mathcal{L}}, \alpha'_0)$ determine the same point in $\mathcal{M}(\widetilde{P})$ if and only if the holomorphic structures on $\mathcal{L}$ are isomorphic and the induced holomorphic isomorphism of $\mathcal{L}_0$ carries the holomorphic sections to constant scalar multiples of each other.

Proof. Let us denote by h the given hermitian metric on $\mathcal{L}$. We begin with a pair $(\overline{\partial}_{\mathcal{L}}, \alpha_0)$ as in the statement of the corollary. By Lemma 7.2.4 there is a hermitian metric h' on $\mathcal{L}$ such, letting A' be the h'-hermitian connection inducing the holomorphic structure $\overline{\partial}_{\mathcal{L}}$, we have

$$F_{A'}^+ = \frac{i}{4}(|\alpha_0|_{h'})^2 \omega.$$

Let $\rho\colon \mathcal{L} \to \mathcal{L}$ be a C^∞ complex linear isomorphism with $\rho^*(h') = h$. Let $A = \rho^*(A')$. Then A is a h-hermitian connection inducing a holomorphic structure on $\mathcal{L}$ which is isomorphic to the holomorphic structure $\overline{\partial}_{\mathcal{L}}$. Clearly, $\alpha = \rho^{-1}(\alpha_0)$ is a holomorphic section of $\mathcal{L}_0$ and

$$F_{A''}^+ = \frac{i}{4}|\alpha|^2 \omega.$$

This proves the first statement of the corollary.

Now let us show that the resulting solution (A, α) is unique up to gauge equivalence. The point is that the function λ which scales the metric is itself unique. This means that the isomorphism ρ in the previous paragraph is unique up to an S^1 change of gauge. From this the uniqueness of (A, α) up to change of gauge is clear.

Notice that if we replace α by a constant complex multiple $\lambda_0 \alpha$ with $\lambda_0 \neq 0$ then the resulting solution to the Seiberg-Witten equations is gauge equivalent to the one produced by α.

By Lemma 7.2.1 any solution to the Seiberg-Witten equations arises in this way.

Lastly, we need to see when two pairs $(\overline{\partial}_{\mathcal{L}}, \alpha_0)$ and $(\overline{\partial}_{\mathcal{L}}', \alpha_0')$ as in the statement of the corollary determine gauge equivalent solutions of the Seiberg-Witten equations. If they determine gauge equivalent solutions, then the holomorphic structures $\overline{\partial}_{\mathcal{L}}$ and $\overline{\partial}_{\mathcal{L}}'$ are isomorphic. Thus, we may assume that $\overline{\partial}_{\mathcal{L}} = \overline{\partial}_{\mathcal{L}}'$. Clearly, if the pairs determine gauge equivalent solutions to the Seiberg-Witten equations then there must be a holomorphic isomorphism of $\overline{\partial}_{\mathcal{L}}$ which sends α_0 to α_0'. But the only holomorphic isomorphisms of a holomorphic line bundle are multiplication by constant complex scalars. This completes the proof. $\square$

There is a completely analogous result when the degree of $\mathcal{L}$ is positive.

Corollary 7.2.6 *Let $\widetilde{P}$ be a Spin^c structure on a Kähler surface X. Suppose that the degree of the determinant $\mathcal{L}$ is positive. Fix a pair $(\overline{\partial}_{\mathcal{L}}, \beta_0)$, where $\overline{\partial}_{\mathcal{L}}$ is a holomorphic structure on $\mathcal{L}$ and $\overline{\beta}_0$ is a non-zero holomorphic section of $K_X \otimes \mathcal{L}_0^{-1} = \sqrt{K_X \otimes \mathcal{L}^{-1}}$. Then there is a solution (A, β)*

to the Seiberg-Witten equations as in Corollary 7.2.3 with the following properties:

- *A determines a holomorphic structure on $\mathcal{L}$ which is isomorphic to the holomorphic structure $\bar{\partial}_{\mathcal{L}}$, and*

- *there is a holomorphic isomorphism from the structure determined by A to $\bar{\partial}_{\mathcal{L}}$ which sends β to β_0.*

Such a solution (A, β) is unique up to gauge equivalence. In this way any pair $(\bar{\partial}_{\mathcal{L}}, \beta_0)$ as above determines a point in the moduli space $\mathcal{M}(\tilde{P})$. All points of $\mathcal{M}(\tilde{P})$ arise in this way. Two pairs determine the same point in $\mathcal{M}(\tilde{P})$ if and only if the holomorphic structures on $\mathcal{L}$ are isomorphic and the induced holomorphic isomorphism of $\mathcal{L}_0$ carries the holomorphic sections to constant scalar multiples of each other.

Lastly, we need to consider the case when the degree of $\mathcal{L}$ is zero. Since the degree of $\mathcal{L}$ is both non-positive and non-negative we conclude that the spinor field (α, β) vanishes identically. This yields the following result.

Corollary 7.2.7 *Let $\tilde{P}$ be a $Spin^c$-structure on a Kähler surface X. If the degree of the determinant line bundle $\mathcal{L}$ of $\tilde{P}$ is zero, then any solution to the Seiberg-Witten equations consists of an anti-self-dual connection A on $\mathcal{L}$ and a trivial spinor field. This identifies the moduli space $\mathcal{M}(\tilde{P})$ with the space of gauge equivalence classes of anti-self-dual connections on $\mathcal{L}$.*

7.3   Evaluation for Kähler Surfaces

First, we explain our convention for orienting $H^2_+(X; \mathbf{R})$ and $H^1(X; \mathbf{R})$ when X is a Kähler surface. Recall that these orientations are necessary in order to give a sign to the Seiberg-Witten invariant. We orient $H^1(X; \mathbf{R})$ in such a way that its projection onto $H^{0,1}(X; \mathbf{C})$ is an orientation-preserving isomorphism when the later space is given the orientation induced from its complex structure. As to $H^2_+(X; \mathbf{R})$, it decomposes as an orthogonal direct sum of the real multiples of the Kähler class and the intersection of $H^{2,0}(X; \mathbf{C}) \oplus H^{0,2}(X; \mathbf{C})$ with $H^2(X; \mathbf{R})$. We orient $H^2_+(X; \mathbf{R})$ by orienting $\mathbf{R}\omega$ so that ω is in the positive subspace, and we orient orthogonal complement so that the projection into $H^{0,2}(X; \mathbf{C})$ is an orientation-preserving isomorphism when the later space is given its complex orientation. From now on when we discuss the Seiberg-Witten invariant of a Kähler surface we implicitly use these orientations.

Proposition 7.3.1 *Let X be a Kähler surface with Kähler metric. Then:*

- *If the degree of K_X is negative then only solutions to the Seiberg-Witten equations are reducible.*

- *Let P_X be the $Spin^c$ structure determined by the complex structure. If the degree of K_X is positive then $SW(P_X) = 1$ when the Seiberg-Witten invariant is computed with respect to the given Kähler metric.*

Proof. Let us first consider the case when K_X has negative degree with respect to the Kähler form. Fix a $Spin^c$ structure $\tilde{P}$. Let us first consider the case when the determinant line bundle $\mathcal{L}$ of $\tilde{P}$ is non-positive. According to Lemma 7.2.1, in this case $\beta = 0$ and α is holomorphic section of $\mathcal{L}_0$. If $\alpha \neq 0$, then we see that the degree of $\mathcal{L}_0$ is ≥ 0. But this is a contradiction since $\mathcal{L}_0^2 = K_X \otimes \mathcal{L}$ clearly has negative degree. It follows that in this case $\alpha = 0$ also and the solution is reducible.

If the degree of $\mathcal{L}$ is non-negative, then according to Lemma 7.2.1 the section $\alpha = 0$ and $\bar{\beta}$ is holomorphic section of $K_X \otimes \mathcal{L}_0^{-1}$. If $\beta \neq 0$, this implies that this bundle has non-negative degree. But this bundle is isomorphic to a square root of $K_X \otimes \mathcal{L}^{-1}$ which implies that it has negative degree. This contradiction shows that $\beta = 0$ and hence in this case as well there are only reducible solutions to the Seiberg-Witten equations. This completes the proof of the first item in the statement of the proposition.

Now let us suppose that the degree of K_X is positive and that the $Spin^c$-structure $\tilde{P}_X$ that we are considering is the one induced by the complex structure. Of course the determinant line bundle $\mathcal{L}$ is equal to K_X^{-1} and hence has negative degree. This is the $Spin^c$-structure for which $\mathcal{L}_0$ is trivial as a C^∞ complex line bundle. Suppose that we have a solution (A, α) to the Seiberg-Witten equations for $\tilde{P}_X$. The holomorphic structure on $\mathcal{L}_0$ induced by A_0 has a non-trivial holomorphic section α. Since the bundle is topologically trivial, this holomorphic section must be nonwhere zero and α is a constant section. This proves that the moduli space $\mathcal{M}(\tilde{P}_X)$ is a single point.

Next we show that this point is a smooth point by computing the differentials in the elliptic complex associated to the solution (A, α). The complex is

$$0 \longrightarrow \Omega^0(X; i\mathbf{R}) \xrightarrow{D_1} \begin{array}{c} \Omega^1(X; i\mathbf{R}) \\ \oplus \\ \Omega^0(X; \mathbf{C}) \oplus \Omega^{0,2}(X; \mathbf{C}) \end{array} \xrightarrow{D_2} \begin{array}{c} \Omega_+^2(X; i\mathbf{R}) \\ \oplus \\ \Omega^{0,1}(X; \mathbf{C}) \end{array} \longrightarrow 0$$

$$(7.7)$$

where

$$D_1(if) = \begin{pmatrix} 2idf \\ -if \cdot \alpha \end{pmatrix}$$

$$D_2 \begin{pmatrix} i\lambda \\ (a,b) \end{pmatrix} = \begin{pmatrix} P_+ d(i\lambda) - (i\mathrm{Re}(a\bar{\alpha})/2)\,\omega + (\alpha\bar{b} - \bar{\alpha}b)\,/2 \\ \sqrt{2}\bar{\partial}(a) + \sqrt{2}\bar{\partial}^*(b) + \pi^{0,1}(i\lambda) \cdot \alpha/\sqrt{2} \end{pmatrix}$$

Since α is a constant, non-zero section, it is clear that the kernel of D_1 is trivial. Let us consider the kernel of D_2. Suppose that

$$D_2 \begin{pmatrix} i\lambda \\ (a,b) \end{pmatrix} = 0.$$

Applying $\bar{\partial}$ to the second coordinate of $D_2 \begin{pmatrix} i\lambda \\ (a,b) \end{pmatrix}$ and using the fact that $\bar{\partial} \circ \bar{\partial} = 0$ and that $\bar{\partial}\alpha = 0$, we conclude that

$$\frac{1}{2}\bar{\partial}\left(\pi^{0,1}(i\lambda)\right) \cdot \alpha + \bar{\partial}\bar{\partial}^*(b) = 0.$$

We also have

$$\bar{\partial}\left(\pi^{0,1}(i\lambda)\right) = d(i\lambda)^{0,2} = \frac{\bar{\alpha}b}{2}.$$

(The second equality uses the fact that the first coordinate of $D_2 \begin{pmatrix} i\lambda \\ (a,b) \end{pmatrix}$ is zero.) Plugging this in gives

$$\frac{1}{4}|\alpha|^2 b + \bar{\partial}\bar{\partial}^* b = 0.$$

Taking the L^2-inner product with b we find that

$$\frac{1}{4}\|\bar{\alpha}b\|_{L^2}^2 + \|\bar{\partial}^*(b)\|_{L^2}^2 = 0,$$

and hence $\bar{\alpha}b = 0$, implying that $b = 0$.

We write $i\lambda = \bar{\xi} - \xi$ for some $\bar{\xi} \in \Omega^{0,1}(X; \mathbf{C})$. The equations telling us that the element is in the kernel of D_2 now become

$$\sqrt{2}\bar{\partial}a + \frac{1}{\sqrt{2}}\bar{\xi} \cdot \alpha = 0 \tag{7.8}$$

$$P_+(\partial\bar{\xi} - \bar{\partial}\xi) = \frac{i}{2}\mathrm{Re}(a\bar{\alpha})\omega. \tag{7.9}$$

We write $a = (u + iv)\alpha$ with u and v being real-valued functions. By adding $D_1(iv)$ to $(i\lambda, a, b)$ we arrange that in fact $a = u\alpha$ with u a real-valued function. Equation 7.8 now reads

$$\sqrt{2}\bar{\partial}u \cdot \alpha + \frac{1}{\sqrt{2}}\bar{\xi} \cdot \alpha = 0$$

from which we conclude

$$\overline{\xi} = -2\overline{\partial}u.$$

Using this Equation 7.9 is equivalent to

$$8\Delta(u) + u = 0.$$

Since Δ has a non-negative spectrum, this implies that $u = 0$. We conclude that $i\lambda = 0$ and that $\alpha = 0$. This proves that any element in the kernel of D_2 is in the image of D_1 and hence that the first cohomology of the elliptic complex is trivial.

Lastly, we need to compute the second cohomology of the elliptic complex. But we know that the index of the complex is given by

$$\left(K_X^2 - (2\chi(X) + 3\sigma(X))\right)/4.$$

Since X is a Kähler manifold, it follows that this index is zero. Since $H^0 = H^1 = 0$, it follows that the second cohomology is zero as well. This completes the proof that the unique solution to the equations is a smooth point of the moduli space and hence that the Seiberg-Witten invariant of the $Spin^c$-structure $\widetilde{P}_X$ is ± 1.

Our next task is to determine the sign. We decompose the operators $D_1 = \overline{D}_1 + E_1 = 2d + E_1$ and $D_2 = \overline{D}_2 + E_2 = (P_+d + \partial\!\!\!/_A) + E_2$, with E_1 and E_2 being operators of order zero. We decompose the spaces in the elliptic complex into $\mathcal{H}^i$ and $(\mathcal{H}^i)^\perp$, where $\mathcal{H}^i$ is the space of harmonic forms for the complex made using the $\overline{D}_i$ instead of the D_i, and $(\mathcal{H}^i)^\perp$ is the L^2-orthogonal subspace. It is easy to see that the operators E_i preserve this decomposition. By construction, the orientation for the determinant line bundle of the elliptic complex is given by the orientations for the spaces $\mathcal{H}^i$ induced by the given orientations on $H^i(X; i\mathbf{R})$ and the complex orientations on the spaces of harmonic spinors. To compute the sign of the determinant line bundle at the solution we need only consider the sign of the determinant line bundle of the following finite dimensional complex

$$0 \to H^0(X; i\mathbf{R}) \xrightarrow{E_1} \begin{array}{c} H^1(X; i\mathbf{R}) \\ \oplus \\ H^0(X; \mathbf{C}) \oplus H^{0,2}(X; \mathbf{C}) \end{array} \xrightarrow{E_2} \begin{array}{c} H^2_+(X; i\mathbf{R}) \\ \oplus \\ H^{0,1}(X; \mathbf{C}) \end{array} \to 0$$

when the complex cohomology groups are given their complex orientations and the cohomology groups with coefficients $i\mathbf{R}$ are given the chosen orientations. Clearly, with these orientations, the map induced by E_1 from $H^1(X; i\mathbf{R}) \to H^{0,1}(X; \mathbf{C})$ is an orientation-preserving isomorphism. Likewise, the map induced by E_1 from $H^{0,2}(X; \mathbf{C})$ to $H^2_+(X; i\mathbf{R})$ is an orientation-preserving isomorphism onto the orthogonal complement of the multiples of the Kähler class. Factoring out these isomorphisms we are left with

$$0 \to i\mathbf{R} \to \mathbf{C} \to i\mathbf{R} \to 0$$

where the first map is minus the inclusion and the second map is $\frac{i}{2}$ times the real part. Clearly, with the given orientations the sign of determinant of this exact complex is $+1$. This completes the proof that $SW(\tilde{P}_X) = +1$.

□

7.4   Computation for Kähler Surfaces

Theorem 7.4.1 *Let X be a minimal algebraic surface of general type. Then for any Kähler metric we have*

$$SW(\tilde{P}) = \begin{cases} 1 & \text{if } \tilde{P} \cong \tilde{P}_X \\ (-1)^{1+p_g(X)-q(X)} & \text{if } \tilde{P} \cong -\tilde{P}_X \\ 0 & \text{otherwise.} \end{cases}$$

Proof.   A minimal surface of general type is characterized by the fact that $K_X^2 > 0$ and that K_X is numerically effective; i.e., for any effective divisor $D \subset X$ we have $K_X \cdot D \geq 0$. Let ω be the Kähler form of a Kähler metric on X. It follows from the numerically effective condition that $K_X \cdot \omega \geq 0$. If $K_X \cdot \omega = 0$ then by the Hodge index theorem, it follows that $K_X^2 \leq 0$. This is impossible, so that we see that the degree of K_X with respect to ω is positive. The evaluation of $SW(\tilde{P}_X)$ is now immediate from Proposition 7.3.1. The evaluation of $SW(-\tilde{P}_X)$ is then a consequence of Corollary 6.8.4.

Let us show that these are the only two *Spin^c* structures on which SW is non-zero. Let $\tilde{P}$ be a *Spin^c* structure with non-negative formal dimension. This means that its determinant line bundle $\mathcal{L}$ satisfies $c_1(\mathcal{L})^2 \geq K_X^2 > 0$. In particular, $c_1(\mathcal{L})^+$ is not a torsion class. This means that there are no reducible solutions, and hence the degree of $\mathcal{L}$ is non-zero. By the symmetry of SW under the involution, we can assume that the degree of $\mathcal{L}$ is negative. We wish to show that if $SW(\tilde{P}) \neq 0$, then $\tilde{P} \cong \tilde{P}_X$. Of course, if $SW(\tilde{P}) \neq 0$, then there is a solution to the Seiberg-Witten equations for $\tilde{P}$.

A solution implies the existence of a holomorphic structure on $\mathcal{L}$ for which $\mathcal{L}_0 = \sqrt{K_X \otimes \mathcal{L}}$ has a non-zero holomorphic section. Let $L = c_1(\mathcal{L})$. Invoking the numerically effective criterion again and the fact that $\mathcal{L}_0$ has a non-trivial holomorphic section, we conclude that $K_X \cdot (K_X + L)/2 \geq 0$. Since K_X has positive degree and $\mathcal{L}$ has negative degree, there is $t \geq 0$ such that

$$\omega \cdot (K_X + tL) = 0.$$

By the Hodge index theorem this implies that

$$0 \geq (K_X + tL)^2 = K_X^2 + 2tK_X \cdot L + t^2 L^2.$$

The minimum value of this quadratic function of t occurs when $t = -(K_X \cdot L)/L^2$ and is equal to

$$K_X^2 - \frac{K_X \cdot L}{L^2}.$$

Since $L^2 \geq K_X^2 \geq -K_X \cdot L$ and $K_X^2 > 0$, we see that this quantity is non-negative, and hence must be equal to zero. This means that the only non-positive value for the above expression as a function of t is when $t = 1$. Hence, $(K_X + L)^2 = 0$ and $\omega \cdot (K_X + L) = 0$. By the Hodge index theorem these two equalities imply that $K_X + L$ is a torsion cohomology class, or equivalently that $c_1(\mathcal{L}_0)$ is a torsion class. Since $\mathcal{L}_0$ has a non-trivial holomorphic section, this implies that $\mathcal{L}_0$ is holomorphically trivial and hence that the $Spin^c$-structure in question is $\widetilde{P}_X$. $\qquad\square$

Corollary 7.4.2 *If X is a minimal algebraic surface of general type then the Seiberg-Witten function is independent of the choice of Kähler metric.*

Of course, this is a new statement only in the case when $b_2^+(X) = 1$.

We have an immediate consequence for the differential topology of Kähler surfaces.

Corollary 7.4.3 *Suppose that X and Y are minimal Kähler surfaces of general type and that $f\colon X \to Y$ is an orientation-preserving diffeomorphism. Then $f^*(K_Y) = \pm K_X$.*

Now let us consider the case when X is an elliptic surface.

Theorem 7.4.4 *Suppose that X is a minimal Kähler surface which is elliptic and for which K_X is not a torsion class. Then*

$$SW(\widetilde{P}_X) = 1$$

$$SW(-\widetilde{P}_X) = (-1)^{1 + p_g(X) - q(X)}.$$

Furthermore, if $\widetilde{P}$ is a $Spin^c$-structure for X with $SW(\widetilde{P}) \neq 0$ then the image of $c_1(\mathcal{L})$ in rational cohomology is a rational multiple of the image of K_X, with the multiple being between 1 and -1. If the multiple is ± 1, then the $Spin^c$-structure is $\pm \widetilde{P}_X$.

Proof.   For all surfaces listed in this theorem we have that K_X is numerically effective and $K_X^2 = 0$. Thus, since K_X is numerically effective it follows that $K_X \cdot \omega \geq 0$. Since K_X is assumed not to be a torsion class, it follows from the Hodge index theorem that $K_X \cdot \omega \neq 0$. hence K_X has positive degree, and the computation for $SW(\tilde{P}_X)$ and $SW(-\tilde{P}_X)$ follows from Corollaries 7.3.1 and 6.8.4.

Now let us suppose that we have a $Spin^c$-structure $\tilde{P}$ for which the Seiberg-Witten invariant is non-zero. The properties that we shall use are that there are solutions to the Seiberg-Witten equations for $\tilde{P}$ and that the formal dimension of the moduli space is non-negative. By symmetry under the involution, it suffices to consider the case when the determinant line bundle $\mathcal{L}$ is of non-positive degree. The dimension condition implies that $\mathcal{L}^2 \geq K_X^2 = 0$. Hence, we see that if the degree of $\mathcal{L}$ is zero then $c_1(\mathcal{L})^2 = 0$ and hence $c_1(\mathcal{L})$ is a torsion class. This is allowed in the proposition.

From now on we assume that the degree of $\mathcal{L}$ is negative. In particular, the holomorphic section α of $\mathcal{L}_0$ is non-trivial. As before, it follows from this fact and the numerical effectiveness of K_X that $K_X \cdot (K_X + L) = K_X \cdot L \geq 0$. For some $t_0 > 0$ we have that the degree of $K_X + t_0 L$ is zero, and hence by the Hodge index theorem

$$2t_0 K_X \cdot L + t_0^2 L^2 \leq 0$$

and is equal to zero if and only if $K_X + t_0 L$ is a torsion class. Since $K_X \cdot L \geq 0$, it follows that $L^2 \leq 0$. Let us show that $L^2 = 0$. For suppose not. Then the minimum of the previous expression, viewed as a function of a real-variable t, occurs at $t = -(K_X \cdot L)/L^2$ and is equal to

$$\frac{-(K_X \cdot L)^2}{L^2}.$$

Since $L^2 < 0$, this expression is positive, yielding a contradiction. Thus, we conclude that $L^2 = 0$ and hence $K_X \cdot L = 0$. This means that $K_X + t_0 L$ is a torsion class. Since K_X and L are integral cohomology classes with K_X non-trivial modulo torsion, it must be the case that t_0 is rational.

This multiple t_0 can not be less than -1 or else, $\mathcal{L}_0$ would be of negative degree, contradicting the fact that it has a non-trivial holomorphic section. In the same vein as before, if the rational multiple is -1, then $\mathcal{L}_0$ would be of degree zero and have a non-trivial holomorphic section implying that it is holomorphically trivial. This means that the $Spin^c$-structure in question is $\tilde{P}_X$.

The case when $\mathcal{L}$ has positive degree follows by symmetry. $\square$

Corollary 7.4.5 *Suppose that X and Y are minimal Kähler surfaces each of which is either of general type or is an elliptic surface with canonical class which is not a torsion cohomology class. Suppose that $f: X \to Y$ is an orientation-preserving diffeomorphism. Then $f^*(K_Y) = \pm K_X$.*

Note. It is possible to prove analogous results when K_X is a torsion class, for example for the $K3$ surface, but this requires a more detailed study of reducible solutions. See [3] for a description of the argument in these cases.

We finish this section with a discussion of the case of blow ups of the surfaces covered by the last two theorems.

Theorem 7.4.6 *Suppose that X is a Kähler surface whose minimal model $\overline{X}$ is either a surface of general type or an elliptic surface with $K_{\overline{X}}$ not a torsion class. Suppose that the nontrivial fibers of $X \to \overline{X}$ are exceptional curves $E_1, \ldots E_k$. Let the Kähler metric have a Kähler class ω of the form*

$$\omega = \overline{\omega} + \sum_{i=1}^{k} \epsilon_i [E_i]^*$$

where the ϵ_i are sufficiently small positive real numbers and $[E_i]^*$ is the cohomology class Poincaré dual to the ith exceptional curve. Suppose that $\widetilde{P}$ is a Spin^c structure on X for which the Seiberg-Witten invariant is non-trivial. Then $\widetilde{P}$ is isomorphic to the tensor product of the pullback of a Spin^c structure $\overline{P}$ on $\overline{X}$ which has non-zero Seiberg-Witten invariant with a $U(1)$-bundle whose first Chern class is of the form $\pm E_1 \pm \cdots \pm E_k$. Furthermore,

$$SW_X(\widetilde{P}) = \pm SW_{\overline{X}}(\overline{P}).$$

Proof. Let $\pi: X \to \overline{X}$ be the projection to the minimal model. Suppose that $\mathcal{L}_0$ is a bundle with a non-trivial holomorphic section α. Then the pushforward $\overline{\mathcal{L}}_0 = \pi_*(\mathcal{L}_0)$ is a line bundle on $\overline{X}$ and $\overline{\alpha} = \pi_*(\alpha)$ is a non-trivial holomorphic section. Thus, we see that if $\widetilde{P}$ is any Spin^c structure on X whose determinant line bundle $\mathcal{L}$ has negative degree and for which there is a solution to the Seiberg-Witten equations, then the pushforward $\overline{P}$ is a Spin^c structure on $\overline{X}$ whose determinant line bundle $\overline{\mathcal{L}}$ has negative degree and for which the Seiberg-Witten equations have a solution.

The first Chern class $c_1(\mathcal{L})$ is of the form $\pi^*(c_1(\overline{\mathcal{L}})) + \sum_{i=1}^{k} n_i [E_i]^*$ where the n_i are odd integers. It follows that

$$c_1(\mathcal{L})^2 - K_X^2 \leq c_1(\overline{\mathcal{L}})^2 - K_{\overline{X}}^2.$$

Since we have already seen that for all *Spin^c*-structures $\overline{P}$ on $\overline{X}$ for which the Seiberg-Witten equations have solutions have the property that

$$c_1(\overline{\mathcal{L}})^2 \leq K_{\overline{X}}^2.$$

Thus, the same is true on X and the only solutions for which the formal dimension is non-negative occur for *Spin^c* structures whose determinant line bundles $\mathcal{L}$ satisfy

$$c_1(\mathcal{L}) = \pi^* c_1(\overline{\mathcal{L}_l}) + \sum_{i=1}^{k} \pm [E_i]^*$$

where $\overline{\mathcal{L}}$ is the determinant line bundle of a *Spin^c*-structure on $\overline{X}$ for which there is a solution to the Seiberg-Witten equations with non-negative index.

We leave the identification of the Seiberg-Witten solutions and the resulting identification of the Seiberg-Witten invariants to the reader. (See [3] for more details.) $\square$

7.5   Final Remarks

We have concentrated in this chapter on solving the Seiberg-Witten equations over Kähler manifolds. There is a series of recent papers by Taubes [12, 13, 14] where he establishes (somewhat weaker) analogues of these results for symplectic manifolds. In particular, Taubes shows that the value of the Seiberg-Witten invariant of a symplectic manifold with $b_2^+ > 1$ on the inverse of the canonical class of the symplectic structure is always equal to 1, generalizing our result here for Kähler manifolds. He goes on to establish an analogue of Theorem 7.4.4 and also to show that the Seiberg-Witten invariant is equal to the Gromov invariant counting the number of pseudo-holomorphic curves (passing through a fixed number of points). These results are surely just the beginning of a vast program of finding the appropriate symplectic analogues of the standard results in Kähler geometry. Time will tell how strong the analogies will be, but it is clearly a promising area for further research.

Bibliography

[1] S. Donaldson and P. Kronheimer, 'The Geometry of Four-Manifolds,' Clarendon, Oxford, 1990.

[2] R. Friedman and J. Morgan, 'Smooth Four-Manifolds and Complex Surfaces,' in Ergebnisse der Mathematik und ihrer Grensgebiete 3. Folge, Band 27, Springer-Verlag, New York, 1994.

[3] R. Friedman and J. Morgan, Algebraic surfaces and Seiberg-Witten Invariants, (to appear).

[4] P. Griffiths and J. Harris, 'Principles of Algebraic Geometry,' Wiley, New York, 1978.

[5] J. Kazdan and F. Warner, Curvature functions for compact 2-manifolds, Ann. of Math. **99** (1974), 14-47.

[6] P. Kronheimer and T. Mrowka, Gauge theory for embedded surfaces I and II, (to appear).

[7] B. Lawson and M-L. Michelson, 'Spin Geometry,' Princeton University Press, Princeton, 1989.

[8] J. Morgan, Z. Szabó and C. Taubes, A Product Formula for the Seiberg-Witten Invariants and the Generalized Thom Conjecture, (to appear).

[9] N. Seiberg and E. Witten, Electromagnetic duality. monopole condensation and confinement in N=2 supersymmetric Yang-Mills theory, Nucl. Phys. **B426** (1994), 19-52.

[10] N. Seiberg and E. Witten, Monopoles, duality and chiral symmetry breaking in N=2 supersymmetric QCD, Nucl. Phys. **B 431** (1994), 581-640.

[11] C. Taubes, Self-dual connections on non-self-dual four-manifolds, J. Differ. Geom. **17** (1982) 139-170.

[12] C. Taubes, The Seiberg-Witten invariant and symplectic forms, Math. Res. Letters **1** (1994), 809-822.

[13] C. Taubes, More constraints on symplectic manifolds from Seiberg-Witten equations, Math. Res. Letters **2** (1995), 9-14.

[14] C. Taubes, The Seiberg-Witten and Gromov invariants, Mat. Res. Letters **2** (1995), 221-238.

[15] K. Uhlenbeck, Removable Singularities in Yang-Mills fields, Commun. Math. Phys. **83** (1982), 11-29.

[16] R. Wells, 'Differential Analysis on Complex Manifolds, Prentice Hall, Engelwood Cliffs, N.J. 1973, 2nd ed. Springer-Verlag, New York.

[17] E. Witten, Monopoles and 4-manifolds, Math. Res. Letters **1** (1994) 764-796.